Springer Tracts in Nature-Inspired Computing

The book series is aimed at providing an exchange platform for researchers to summarize the latest research and developments related to nature-inspired computing in the most general sense. It includes analysis of nature-inspired algorithms and techniques, inspiration from natural and biological systems, computational mechanisms and models that imitate them in various fields, and the applications to solve real-world problems in different disciplines. The book series addresses the most recent innovations and developments in nature-inspired computation, algorithms, models and methods, implementation, tools, architectures, frameworks, structures, applications associated with bio-inspired methodologies and other relevant areas.

The book series covers the topics and fields of Nature-Inspired Computing, Bio-inspired Methods, Swarm Intelligence, Computational Intelligence, Evolutionary Computation, Nature-Inspired Algorithms, Neural Computing, Data Mining, Artificial Intelligence, Machine Learning, Theoretical Foundations and Analysis, and Multi-Agent Systems. In addition, case studies, implementation of methods and algorithms as well as applications in a diverse range of areas such as Bioinformatics, Big Data, Computer Science, Signal and Image Processing, Computer Vision, Biomedical and Health Science, Business Planning, Vehicle Routing and others are also an important part of this book series.

The series publishes monographs, edited volumes and selected proceedings.

Khalid Raza
Editor

Solving with Bees

Transformative Applications of Artificial Bee Colony Algorithm

Springer

Editor
Khalid Raza
Department of Computer Science
Jamia Millia Islamia
New Delhi, Delhi, India

ISSN 2524-552X ISSN 2524-5538 (electronic)
Springer Tracts in Nature-Inspired Computing
ISBN 978-981-97-7343-5 ISBN 978-981-97-7344-2 (eBook)
https://doi.org/10.1007/978-981-97-7344-2

This Springer imprint is published by the registered company Springer Nature Singapore Pte Ltd.
The registered company address is: 152 Beach Road, #21-01/04 Gateway East, Singapore 189721, Singapore

This book is dedicated to all those who have inspired and supported me throughout my journey in the realm of Artificial Bee Colony algorithms and their applications.

To my dear parents, whose unwavering love and guidance have been the cornerstone of my achievements. Though you are no longer with me, your memory continues to light my path and inspire my endeavors. Your sacrifices and encouragement have been the driving force behind my pursuit of knowledge and excellence. I miss you deeply and hold your wisdom and love in my heart always.

To my mentors, colleagues, and students, who have shared their insights, offered their support, and engaged in stimulating discussions that have enriched this work.

To the countless researchers and pioneers in the field of nature-inspired algorithms, whose groundbreaking work has laid the foundation for advancements and innovations.

And finally, to my friends and family, whose patience, understanding, and encouragement have sustained me through the highs and lows of this intellectual journey.

Khalid Raza

Preface

In recent years, nature-inspired algorithms have emerged as powerful tools for solving complex optimization problems. Among these, the Artificial Bee Colony (ABC) algorithm has gained significant attention due to its simplicity, flexibility, and effectiveness. The ABC algorithm is a nature-inspired optimization algorithm based on the foraging behavior of honeybees. Introduced by Karaboğa in 2005, the ABC algorithm mimics the intelligent foraging strategies of a honeybee colony to find the optimal solution to complex problems. Honeybees are known for their efficient and sophisticated methods of searching for food sources. The colony is divided into different types of bees, each with a specific role: employed bees, onlooker bees, and scout bees. These bees work together to explore the environment, exploit food sources, and share information. The ABC algorithm's ability to handle complex, multi-dimensional, and nonlinear optimization problems has made it a popular choice among researchers and practitioners. Its inspiration from nature not only provides an effective problem-solving tool but also contributes to the broader field of nature-inspired computation. It has been successfully applied to a wide range of problems, from engineering design to medical diagnosis.

This book, *Solving with Bees: Transformative Applications of Artificial Bee Colony Algorithms* is a comprehensive exploration of the ABC algorithm and its diverse applications. It is the culmination of years of research and collaboration with experts in the field. My goal in editing this book is to provide both a theoretical foundation and practical insights into the application of ABC algorithms. Throughout this book, we strived to present complex concepts in an accessible manner, making it suitable for both beginners and experienced researchers. I hope this work serves as a valuable resource for those looking to understand the potential of ABC algorithms and inspires further research and innovation in this exciting field.

It is my sincere hope that this book will not only enhance your understanding of ABC algorithms but also inspire you to explore new horizons in optimization and problem-solving.

New Delhi, India　　　　Khalid Raza

Acknowledgments

This book is the outcome of countless hours of hard work, dedication, collaboration, and support from many remarkable researchers from different organizations. It is with immense gratitude and humility that I extend my heartfelt thanks to all who have contributed to the creation of *Solving with Bees: Transformative Applications of Artificial Bee Colony Algorithms.*

First and foremost, I would like to express my deepest gratitude to my family. To my parents, whose unwavering love and guidance have been the bedrock of my life, and whose memory continues to inspire me every day. To my spouse (Farhat Meraj) and children (Arham and Hoorain), thank you for your endless patience, understanding, and encouragement throughout this journey. Your support has been my anchor and my motivation.

I am profoundly grateful to my co-authors and contributors, whose expertise and collaboration have enriched this book in numerous ways. To the researchers and pioneers in the field of nature-inspired algorithms, I owe a debt of gratitude. Your groundbreaking work has laid the foundation upon which this book is built. I am also deeply appreciative of my students, whose curiosity and enthusiasm have been a constant source of inspiration. Your questions, ideas, and perspectives have challenged me to think more deeply and to strive for clarity and excellence in my writing and research.

A special note of thanks to the publishing team from Springer. Your professionalism, support, and commitment to excellence have been crucial in bringing this book to life. I am deeply grateful for your support and hard work.

Finally, I would like to thank all the readers and future researchers who will engage with this book. It is my sincere hope that it serves as a valuable resource and inspires further exploration and innovation in the fascinating field of Artificial Bee Colony algorithms.

Khalid Raza

About This Book

The book, *Solving with Bees: Transformative Applications of Artificial Bee Colony Algorithms*, delves into the fascinating world of Artificial Bee Colony (ABC) algorithms, showcasing their potential to solve a wide array of optimization problems. This book is designed to serve as a comprehensive guide for researchers, practitioners, and students interested in nature-inspired algorithms and their applications. It not only covers the theoretical aspects of ABC algorithms but also provides practical insights and case studies, demonstrating their transformative potential in solving real-world problems across various domains.

This book is structured into three comprehensive parts: *Foundations*, *Applications of Artificial Bee Colony*, and *Case Studies*, each contributing to a holistic understanding of ABC algorithms and their potential. The first part of the book, comprising three chapters, lays the groundwork by introducing the fundamental concepts and variants of the ABC algorithm. Chapter "Fundamentals of Artificial Bee Colony Algorithms and Its Variants" provides a detailed overview of the basics and evolution of ABC algorithms. Chapter "The Rise of Artificial Bee Colony Algorithms in Data Science and Machine Learning is Notable" highlights the growing importance of ABC algorithms in data science and machine learning. Chapter "Swarm Intelligence for Optimization: A Bee's-Eye View on Multi-objective and Dynamic Challenges" delves into the use of swarm intelligence, particularly ABC, for tackling multi-objective and dynamic optimization problems. The second part, comprising six chapters, explores the practical applications of ABC algorithms in diverse fields. Chapter "Artificial Bee Colony Algorithms in Control Systems, Robotics and Automation" discusses the integration of ABC in control systems and robotics, while Chapter "Integrating Artificial Bee Colony Algorithms for Deep Learning Model Optimization: A Comprehensive Review" reviews the optimization of deep learning models using ABC. Chapter "An IoMT Enabled Iterative Artificial Bee Colony Approach Using Federated Learning for Detection of Heart Disease" presents a novel application of ABC in healthcare, specifically for heart disease detection using IoMT and federated learning. Chapter "Optimal Design of a Biomedical Amplifier for Minimum Offset Using a Modified ABC Algorithm" focuses on optimizing biomedical amplifier design with a modified ABC algorithm, while Chapter

"Using Honeybees for Gene Expression Profiling: The Artificial Bee Colony Algorithm to Identify Robust Gene Biomarkers for Clinical Diagnosis" explores the use of ABC for gene expression profiling and biomarker identification. Chapter "Smart Diagnostics for Diabetic Retinopathy: Integrating Artificial Bee Colony Algorithms into Medical Image Analysis" illustrates the application of ABC in medical image analysis for diabetic retinopathy diagnosis. The final part, comprising two chapters, presents real-world case studies demonstrating the effectiveness of ABC algorithms in various research areas. Chapter "Artificial Bee Colony Algorithms in Gene Expression Studies: A Case Study" details a case study on the use of ABC in gene expression analysis, while Chapter "Artificial Bee Colony Algorithm in Multi-omics Analysis: A Case Study" discusses the application of ABC in multi-omics data analysis.

Contents

Editor and Contributors

About the Editor

Dr. Khalid Raza is an Associate Professor at the Department of Computer Science, Jamia Millia Islamia (Central University), New Delhi, India. He also served as a ICCR Chair Visiting Professor at the Faculty of Computer and Information Sciences, Ain Shams University, Cairo, Egypt academic session 2017–2018. Dr. Raza has over 14 years of teaching and research experience at undergraduate, postgraduate, and doctoral levels. He has contributed over 140 peer-reviewed articles in refereed international journals, conference proceedings, and book chapters. Dr. Raza has authored/edited 14 books published by reputed publishers like Springer-Nature, Elsevier, and CRC Press. Dr. Raza is an Academic Editor of *PeerJ Computer Science*, Guest Editor of the *npj Precision Oncology* (Nature), *Natural Product Communications* (SAGE), and *Mini-Reviews in Medicinal Chemistry* (Bentham Science). He has an active collaboration with the scientists from leading institutions in India and abroad. He is an Honorary Research Fellow of INIT International University, Malaysia. Dr. Raza has been featured in the list of the World's Top 2% Scientists released by Stanford University (USA) in collaboration with Elsevier for three consecutive years 2021–2023.

Contributors

Faiyaz Ahmad Department of Computer Science, Jamia Millia Islamia, New Delhi, India

Mohammad Ahmad Department of Computer Science, College of Science and Humanities-Dawadmi, Shaqra University, Al Duwadimi, Saudi Arabia

Shaban Ahmad Department of Computer Science, Jamia Millia Islamia, New Delhi, India

Md. Mobin Akhtar Department of Basic Science, College of Applied Medical Sciences, Riyadh Elm University, Riyadh, Saudi Arabia

Faiz Akram Department of Computer Engineering, Faculty of Engineering and Technology, New Delhi, India

Raiyan Ali Department of Biochemistry, All India Institute of Medical Sciences (AIIMS), New Delhi, India

Alisha Ansari School of Computational and Integrative Sciences, Jawaharlal Nehru University, New Delhi, India

Shafaque Aziz Department of Computer Engineering, Faculty of Engineering and Technology, New Delhi, India;
Government Engineering College Samastipur, Bihar, India

Matthew S. Babik Department of Molecular & Integrative Physiology, University of Illinois Urbana-Champaign, Champaign, USA

Nagmi Bano Department of Computer Science, Jamia Millia Islamia, New Delhi, India

Sagarika Choudhury Tejas Networks Ltd., Bengaluru, Karnataka, India

Swagata Devi Faculty of Engineering, Assam Down Town University, Panikhaiti, Guwahati, Assam, India

Syed Akramah Faizi Department of Information Technology, Delhi Technological University, New Delhi, India

Yagyesh Godiyal Department of Computer Application, Galgotias College of Engineering, Greater Noida, Uttar Pradesh, India

Y. Gokul Department of CSE, Vellore Institute of Technology, Vijayawada, Andra Pradesh, India

Saira Hamid Department of Computer Science, Jamia Millia Islamia, New Delhi, India

Mohammed Nazibul Hasan Angstrom Laboratory, FENE Group, Uppsala University, Uppsala, Sweden

Mohammad Nadeem Khalid Department of Electro-Mechanical Engineering Technology, Institute of Applied Technology, Abu Dhabi Polytechnic University, Abu Dhabi, United Arab Emirates

Nayyar Ahmed Khan Department of Computer Science, College of Computing and Information Technology, Shaqra University, Shaqra, Saudi Arabia

S Kumaravel School of Electronics Engineering, Vellore Institute of Technology, Vellore, India

R. S. M. Lakshmi Patibandla Department of CSE, Koneru Lakshmaiah Education Foundation, Guntur, Andra Pradesh, India

Naushad Manzoor Laskar School of Electronics Engineering, Vellore Institute of Technology, Vellore, India

D. Madhusudhana Rao Department of CSE, Amrita Vishwa Vidyapeetham, Guntur, Andra Pradesh, India

Arjun Pandey School of Electronics Engineering, Vellore Institute of Technology, Vellore, India

Sahar Qazi Department of Biochemistry, All India Institute of Medical Sciences (AIIMS), New Delhi, India

Sivaram Rajeyyagari Department of Computer Science, College of Computing and Information Technology, Shaqra University, Shaqra, Saudi Arabia

M. Ramakrishna Murthy Anil Neerukonda Institute of Technology and Sciences, Visakhapatnam, Andhra Pradesh, India

Mritunjay Kr. Ranjan School of Computer Science and Engineering, Sandip University, Nashik, Maharashtra, India

Khan Asif Rashid Department of Computer Science, College of Computing and Information Technology, Shaqra University, Shaqra, Saudi Arabia

Khalid Raza Department of Computer Science, Jamia Millia Islamia, New Delhi, India

Vibha Saroha Department of Computer Application, Galgotias College of Engineering, Greater Noida, Uttar Pradesh, India

Arif Md. Sattar Diwan V.S. Institute of Management Studies, Meerut, India

Shilpi Saxena Department of Computer Application & IT, Lords University, alwar, Rajasthan, India

Ashok Sharma Department of Biochemistry, All India Institute of Medical Sciences (AIIMS), New Delhi, India

Ahmed Masih Uddin Siddiqi Department of Computer Engineering and Application, Mangalayatan University, Aligarh, India

Aditya Narayan Singh Department of Computer Application, Galgotias College of Engineering, Greater Noida, Uttar Pradesh, India

Nripendra Kumar Singh Department of Computer Applications, Galgotias College of Engineering and Technology, Greater Noida, Uttar Pradesh, India

Yash Vardhan Singh Department of Computer Application, Galgotias College of Engineering, Greater Noida, Uttar Pradesh, India

Khushnuma Tanveer Department of Computer Science, Jamia Millia Islamia, New Delhi, India

B. Tarakeswara Rao Kallam Haranadhareddy Institute of Technology, Guntur, Andhra Pradesh, India

Sanjay Kr. Tiwari Department of Mathematics, Magadh University, Bodh-Gaya, India

Shreya Tiwari Department of Computer Science and Engineering, Amity School of Engineering and Technology, Amity University, Noida, India

Foundations

Fundamentals of Artificial Bee Colony Algorithms and Its Variants

Yagyesh Godiyal, Aditya Narayan Singh, Matthew S. Babik, and Nripendra Kumar Singh

1 Introduction

In the past few decades, several optimization algorithms inspired by concepts found in nature have been introduced. Ant colony optimization, evolutionary algorithm, particle swarm optimization, harmony search, and other algorithms are a few instances of this type of method. Due to their broad applicability to various situations, the majority of these algorithms are search strategies based on metaheuristics and are commonly known as multipurpose optimization algorithms. Similar to this, Karaboga first released the Artificial Bee Colony (ABC) method in 2005 as a technical paper addressing numerical optimization issues [1]. The ABC algorithm is based on the foraging behavior of honeybees. It is similar to the natural process through which honeybee colonies search for food sources and exchange information about the quality of these sources [1]. ABC has several significant advantages over other optimization techniques, including:

a. Simplicity, flexibility, and resilience [2].
b. The technique utilizes a smaller number of control parameters in comparison to numerous other search techniques [3].
c. The hybridization with other optimization algorithms is facilitated [2].
d. Proficiency in managing the uncertain expenses associated with a specific goal [4].

Y. Godiyal · A. N. Singh · N. K. Singh (✉)
Department of Computer Applications, Galgotias College of Engineering and Technology, Greater Noida, India
e-mail: nripendra.kumar@galgotiacollege.edu

Y. Godiyal
e-mail: yagyesh.godiyal@galgotiacollege.edu

M. S. Babik
Department of Molecular & Integrative Physiology, University of Illinois Urbana-Champaign, Champaign, USA

K. Raza (ed.), *Solving with Bees*, Springer Tracts in Nature-Inspired Computing,
https://doi.org/10.1007/978-981-97-7344-2_1

e. Simplicity of implementation using fundamental mathematical and logical operations.

ABC algorithms have been widely used across various fields such as engineering [4–6], finance [7, 8], and telecommunications [9, 10]. Researchers have introduced hybrid approaches by using ABC with other metaheuristic techniques or incorporating adaptive strategies to improve the accuracy of algorithm parameters [11, 12]. These advancements aim to address the challenges of balancing exploration and exploitation, enhance solution accuracy, and ensure robust performance across different problem domains. Karaboga and his peers in the last l5 years have provided important work related to ABC algorithms and their applications to different optimization methods [1, 3, 4, 6, 13–15].

The ABC algorithm is also applicable to fields like evolutionary biology and ecology through Approximate Bayesian Computation techniques [16]. Modifying and changing the original ABC algorithm have further expanded its utility and efficiency. For example, a variant designed for constrained optimization problems employs Deb's rules for constraint handling, showcasing improved performance against benchmark problems [17]. Additionally, integrating evolutionary-based similarity search mechanisms into a binary ABC variant for feature selection demonstrates the algorithm's superiority in eliminating irrelevant features, thereby optimizing the feature set for better classification performance [18].

Many different versions of the ABC algorithm have been studied, which shows how flexible it is and how it is always changing. Variants such as the Gbest-guided ABC algorithm and the differential evolution-based ABC/best/1 and ABC/best/2 algorithms provide new ways to solve big data optimization challenges, highlighting the ABC algorithm's scalable nature and its effectiveness in managing complex computational problems [19]. Recent research papers include the application of the ABC algorithm to the electric vehicle routing problem [36], coverage optimization in WSNs [37], and predicting the Water Quality Index [38].

The primary aim of this chapter is to provide a comprehensive overview, although not exhaustive, of the advancements and variations made to the original ABC method. Overall, the goal is to make this chapter a good starting point for researchers and scholars interested in learning ABC concepts and their usage in different areas.

The rest of the chapter is divided as follows: The basic component of the ABC algorithm is explained in Sect. 2. The algorithm and workflow of the ABC algorithm are explained in Sect. 3. Section 4 consists of a mathematical representation of the ABC algorithm along with an update mechanism. The variants of ABC algorithms have been summarized in Sect. 5. The last section contains the conclusion and future scope of ABC algorithms.

2 Key Components of ABC Algorithms

There are three main components of Artificial Bee Colony (ABC) algorithms, namely, employed bees, onlooker bees, and scout bees.

2.1 *Employed Bees*

Employed bees are responsible for exploring potential solutions within the search space. Each employed bee corresponds to a candidate solution, and these bees move around the solution space to find new solutions. The quality of a solution is determined by its fitness, which is evaluated based on the objective function of the optimization problem. Employed bees carry information about the solutions they have explored and share this information with onlooker bees during the subsequent phase.

2.2 *Onlooker Bees*

Onlooker bees make decisions regarding which solutions to focus on for further exploration based on the information provided by employed bees. The probability of an onlooker bee choosing a particular solution is proportional to the quality of that solution. This mechanism reflects the recruitment process observed in real honeybee colonies, where onlooker bees decide which food sources to exploit based on the dances performed by employed bees. Onlooker Bees contribute to the exploitation of promising solutions and help the algorithm converge toward optimal or near-optimal solutions.

2.3 *Scout Bees*

Scout bees play a crucial role in maintaining diversity within the population. If a solution stagnates and does not lead to improvement after a certain number of iterations, the corresponding employed bee becomes a scout bee. Scout bees explore new solutions randomly, introducing diversity into the population and preventing the algorithm from getting stuck in local optima. This exploration–exploitation balance, facilitated by Scout Bees, enhances the algorithm's ability to navigate complex solution spaces and find global optima.

2.4 Role of Food Sources

In the context of Artificial Bee Colony (ABC) algorithms, the terms "food sources" and "solutions" are often used interchangeably to represent the potential solutions within the search space of an optimization problem. The algorithm is inspired by the foraging behavior of honeybees, where the goal is to find the best sources of food. In the ABC algorithm, each potential solution is analogous to a food source, and the quality of a solution is determined by its fitness, which is evaluated based on the objective function of the optimization problem.

2.5 Food Sources (Solutions)

In the algorithm, food sources represent candidate solutions to the optimization problem. Each food source is associated with a position in the solution space, and the quality of a solution is assessed by its fitness, which indicates how well it satisfies the optimization criteria. The algorithm aims to iteratively improve the quality of these solutions by mimicking the foraging behavior of honeybees. Solutions with higher fitness values are considered better, and the objective is to converge toward optimal or near-optimal solutions.

2.6 Role of Food Sources in Employed and Onlooker Bees

Employed bees and onlooker bees interact with food sources during the exploration and exploitation phases. Employed bees explore the solution space by adjusting the position of their associated food source, while onlooker bees make decisions about which food sources to exploit based on the information received from employed bees. The probability of an onlooker bee choosing a particular food source is influenced by the quality (fitness) of that solution.

Scout Bees and the Exploration of New Food Sources: The role of scout bees is to introduce diversity into the population by exploring new and random food sources. If a particular food source (solution) does not lead to improvement after a certain number of iterations, the corresponding employed bee becomes a scout bee. Scout bees explore new positions in the solution space, contributing to the algorithm's ability to escape local optima and maintain a balance between exploration and exploitation.

3 ABC Algorithm's Workflow

The workflow of the ABC algorithm illustrated in Fig. 1 can be summarized in the following steps:

- **Initialization**: The algorithm starts by randomly generating a population of solutions, where each solution represents a food source and its quality corresponds to the fitness of the solution (the objective function value).
- **Employed Bees Phase**: Each employed bee is associated with one food source. They explore new food sources (new solutions) near their current food source and evaluate their fitness. If the new food source is better than the current one, the bee memorizes the new location and forgets the old one.
- **Onlooker Bees Phase**: Onlooker bees select food sources based on the probabilities related to their fitness. The better the food source, the higher the chance it will be chosen by an onlooker bee. Once a food source is chosen, the onlooker bee performs a search around that food source, similar to the employed bees.
- **Scout Bees Phase**: If a food source is not improved upon for a predetermined number of cycles (abandoned), it is assumed to be exhausted, and a scout bee is sent out to randomly find a new food source, replacing the abandoned one.
- **Memorization**: At the end of each cycle, the bees memorize the best food source found so far.

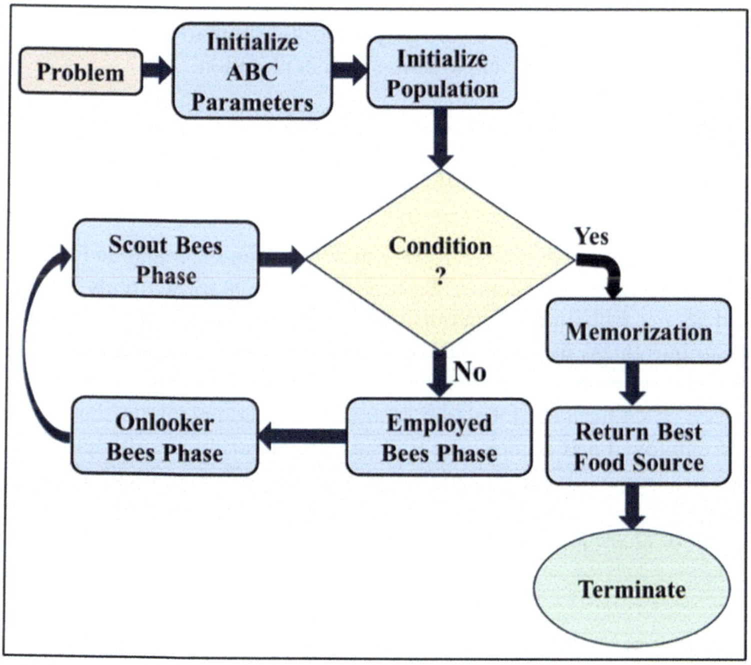

Fig. 1 The workflow of the artificial bee colony (ABC) algorithm

- **Termination**: The algorithm repeats steps 2–5 until a termination criterion is met, such as reaching a maximum number of cycles or achieving a satisfactory solution.

3.1 How Employed Bees Explore the Search Space

Employed bees explore the search space by investigating the vicinity of their current food sources. They employ a local search strategy where they modify the position of their food source using a formula that considers the position of their current food source and the position of a randomly chosen food source from the neighborhood. This method allows the algorithm to exploit the search space effectively by intensifying the search around promising solutions.

3.2 Onlooker Bees' Role in Selecting Promising Solutions

Onlooker bees wait on the dance floor (within the hive) and choose food sources based on the waggle dances performed by the employed bees, which communicate the location and quality of food sources. The selection process is probabilistic and biased towards better food sources, ensuring that higher-quality solutions receive more attention. This selective pressure guides the algorithm towards promising areas of the search space, balancing exploitation with exploration.

3.3 Handling of Abandoned Solutions by Scout Bees

Scout bees come into play when a food source has been exploited and is no longer improving, deemed as abandoned. A scout bee's role is to randomly search for a new food source without any reference to the current food source locations. This mechanism introduces diversity into the population, preventing premature convergence on local optima by exploring new and potentially unvisited regions of the search space. It's a form of global search strategy that complements the local search efforts of employed and onlooker bees. The ABC algorithm is efficient because it combines local search methods (employed and onlooker bees) with global search strategies (scout bees), allowing it to navigate complex search spaces effectively and find optimal or near-optimal solutions to optimization problems.

4 Mathematical Representation of the ABC Algorithm

The ABC algorithm can be mathematically represented through a set of equations that describe the key components and processes involved in the optimization. Here is a simplified mathematical representation:

Initialize parameters:
$N =$ *number of food sources (solutions)*
$D =$ *number of parameters (dimensions) to optimize*
Limit $=$ *maximum number of trials before abandoning a food source*
MaxCycles $=$ *maximum number of cycles (iterations)*
Initialize population:
for $i = 1$ to SN do
for $j = 1$ to D do
$X_i[j] = X_{\min}[j] + \text{rand}() \times (X_{\max}[j] - X_{\min}[j])$
$\text{EvaluateFitness}(X_i)$
Begin optimization loop:
for cycle $= 1 to MaxCyclesdo$
for $i = 1$ to SN do
$k = \text{randIndex}(SN, i)$ // Random index different from i
$j = \text{randIndex}(D)$ // Random dimension
$\Phi = \text{rand}(-1,1)$ // Random perturbation coefficient
$V_i[j] = X_i[j] + \Phi \times (X_i[j] - X_k[j])$ // Generate new solution
$\text{EvaluateFitness}(V_i)$
if $\text{Fitness}(V_i) > \text{Fitness}(X_i)$
$X_i = V_i$ // Greedy selection
// *Calculate probabilities for onlooker bees*
$$\text{Sum Fitness} = \sum_{i=1}^{SN} \text{Fitness}(X_i)$$
for $i = 1$ to SN do
$$P_i = \frac{\text{Fitness}(X_i)}{\text{Sum Fitness}}$$
// *Onlooker Bee Phase*
for $i = 1$ to SN do
if $\text{rand}(0,1) < P_i$
$k = \text{randIndex}(SN, i)$
$j = \text{randIndex}(D)$
$\Phi = \text{rand}(-1,1)$
$V_i[j] = X_i[j] + \Phi \times (X_i[j] - X_k[j])$
$\text{EvaluateFitness}(V_i)$
if $\text{Fitness}(V_i) > \text{Fitness}(X_i)$
$X_i = V_i$ // Greedy selection
// *Scout Bee Phase*
for $i = 1$ to SN do
if $\text{Trials}[i] > \text{Limit}$
for $j = 1$ to D do

$X_i[j] = X_{\min}[j] + \text{rand}() \times (X_{\max}[j] - X_{\min}[j])$
EvaluateFitness(X_i)
Trials[i] = 0
Memorize the best solution found
// End of optimization loop
Output the best solution
END.

4.1 Step-by-Step Discussion of the ABC Algorithm

Step 1: Initialization:

Initialize the population of employed bees, onlooker bees, and scout bees. Generate initial solutions for each employed bee.

Step 2: Objective Function:

Define the objective function to be optimized. Let $f(x)$ represents the objective function, where x is the solution vector.

Step 3: Employed Bee Phase:

- Each employed bee i produces a new solution x_i' by exploring the neighborhood of its current solution x_i using a perturbation process.
- Evaluate the fitness of the new solution:

$$fitness_i' = f(x_i') \tag{1}$$

- If $fitness_i' < fitness_i$, replace x_i with x_i'.

Step 4: Onlooker Bee Phase:

- Select onlooker bees based on the fitness values of employed bees.
- An onlooker bee j chooses an employed bee i with a probability proportional to its fitness:

$$P_{ij} = \frac{fitness_i}{\sum_{k=1}^{n} fitness_k} \tag{2}$$

- The selected onlooker bee generates a new solution x_j' using the perturbation process similar to the employed bee phase.
- Evaluate the fitness of the new solution:

$$fitness_j' = f(x_j') \tag{3}$$

- If $fitness_j' < fitness_j$, replace x_j with x_j'.

Step 5: Scout Bee Phase:

- If an employed bee or an onlooker bee cannot improve its solution after a certain number of iterations, it becomes a scout bee.
- The scout bee generates a new random solution to explore a different region.

Step 6: Termination Criteria:

- Repeat the employed bee and onlooker bee phases until a termination criterion is met (e.g., a maximum number of iterations or a satisfactory solution is found).

4.2 Update Mechanism for Employed and Onlooker Bees

The ABC algorithm utilizes specific equations to update the positions of the employed and onlooker bees in the search space, effectively guiding the search for optimal solutions. These update mechanisms are crucial for exploring the vicinity of current solutions (food sources) and exploiting promising areas of the search space. Here are the key equations and concepts used by employed and onlooker bees:

Employed Bees

The position update for an employed bee is given by the Eq. (4):

$$v_{ij} = x_{ij} + \phi_{ij}.(x_{ij} - x_{kj}) \tag{4}$$

where, v_{ij} is the new position (solution) for the ith bee on the jth dimension.

x_{ij} is the current position of the ith bee on the jth dimension.

x_{kj} is the position of a randomly selected bee (k) different from i on the jth dimension. ϕ_{kj} is a random number between $[-1, 1]$, controlling the step size of the position change. It determines the direction and magnitude of the search in the dimension j.

This equation allows the employed bee to explore new solutions near its current solution by adjusting its position towards or away from another randomly chosen solution. The randomness introduced ensures diversity in the search, allowing the algorithm to explore the search space effectively.

Onlooker Bees

The onlooker bees select a food source based on a probability related to its fitness and then use the same equation as employed bees to update their position. The probability p_i of selecting the ith food source is given by Eq. (5):

$$p_i = \frac{fit_i}{\sum_{n=1}^{SN} fit_n} \tag{5}$$

where, p_i is the probability of selecting the ith food source. fit_i is the fitness of the ith food source. SN is the total number of food sources.

The fitness fit_i usually depends on the objective function that the algorithm is trying to optimize. For maximization problems, the fitness can be directly proportional to the objective function value, while for minimization problems, a transformation may be required to ensure that better solutions have higher fitness values.

5 Variants of ABC Algorithm

Several variants of the ABC algorithm have been proposed to improve its performance across different types of optimization problems. These variants often aim to enhance the exploration and exploitation abilities of the algorithm, balance the trade-off between these two aspects, or adapt it to specific problem domains.

5.1 Gbest-Guided Artificial Bee Colony (GABC) Algorithm

The Gbest-guided Artificial Bee Colony (GABC) Algorithm is a version that enhances the normal ABC algorithm by incorporating the concept of the global best solution. This concept guides the search process and improves the algorithm's capacity to exploit optimal solutions. In their study, Sharma et al. [20] provide an enhanced version of the Gbest-guided artificial bee colony method, designed to increase its performance in solving constrained optimization issues. The changes made to the position update phase in both the employed and onlooker bee phases introduce variation in the search area and strengthen the identification of certain places. The new variant underwent evaluation using 21 benchmark optimization problems, demonstrating its viability as a potential substitute for the original Gbest-guided ABC in constraint optimization issues. Notably, it exhibited enhancements in efficiency and the capacity to discover optimal solutions. The study conducted by Bhambu et al. [21] presents a modified version of the Gbest-guided ABC algorithm called modified GABC (MGABC). The objective of this modification is to enhance the effectiveness of solving constrained optimization issues. MGABC iteratively modifies the step size of solutions while searching for global optima. MGABC's effectiveness is proven by conducting tests on eight real-world complicated optimization issues. The results show that it outperforms ABC, best-so-far ABC, and GABC, making it a strong candidate for addressing both limited and unconstrained optimization problems. The Gbest-guided ABC algorithm is improved by Sharma et al. [22] with the integration of a "Fully Informed Learning" method during the onlooker bee phase. This enhancement results in the creation of the fully informed ABC (FABC) algorithm. FABC seeks to address the problem of being stuck in local optima by utilizing data from every individual in the population. The algorithm's performance was assessed on 20 benchmark optimization problems and it shows promising results. Performance was also compared to other recent variations of ABC and traditional optimization techniques to determine the effectiveness of the proposed method.

5.2 *Modified Artificial Bee Colony (MABC) Algorithm*

The Modified Artificial Bee Colony (MABC) Algorithm is a variant of the ABC algorithm. In this variant, the focus is to alter the procedure for selecting food sources or introducing new features to prevent becoming stuck in suboptimal solutions and improve the exploration process.

Wang et al. [23] present an enhanced version of the ABC algorithm that utilizes chaotic mapping to increase the variety within the bee population. This modification directs the search for potential solutions towards more viable paths, resulting in increased exploitation. The update also involves implementing a worldwide solution during the Scout stage to avoid local optima. The enhanced ABC algorithm was more accurate in comparison to alternative approaches.

The study by Kiran and Findik [24] proposes an enhancement to the ABC algorithm by integrating directional information for artificial bees. The objective is to enhance the system's ability to converge towards global or near-global optima. The suggested approach improves the quality and reliability of solutions for nine widely recognized numerical benchmark functions, demonstrating its effectiveness and superiority compared to the basic ABC and additional variants that incorporate a control parameter (modification rate-MR).

5.3 *Binary Artificial Bee Colony (BABC) Algorithm*

Designed for optimization problems where the decision variables are binary, this variant modifies the solution update equations to operate in binary search spaces.

Santana et al. [25] introduce a novel approach named NBABC, featuring a mechanism that limits the number of dimensions changeable in the employed and onlooker bees' phase. This study compares NBABC with other binary ABC variants and several well-known methods across different tasks, including the OneMax problem, variations of the 0/1 Knapsack problems, and Feature Selection using public datasets. The results demonstrate NBABC's competitive performance, even outperforming other algorithms in certain cases.

Durgut [26] modifies the ABC algorithm for binary optimization problems, naming the variant improved binary ABC (IbinABC). This study focuses on enhancing the ABC algorithm's exploration ability to avoid local minima, employing a fitness value-based update mechanism and varying decision variable selection. The IbinABC algorithm's performance demonstrated on the incapacitated facility location problem using the OR-Library dataset, shows superiority in convergence speed and robustness compared to other algorithms, highlighting its effective binary optimization capability.

Ozturk et al. [15] propose an improved version of the binary artificial bee colony algorithm (IDisABC) for dynamic clustering, tackling the challenge of the discrete artificial bee colony's limited solution generation mechanism. By enhancing solution generation through genetically inspired components, IDisABC demonstrates superiority in dynamic clustering over classical approaches like K-means, and FCM, and other evolutionary computation algorithms. This showcases the algorithm's ability to handle one of the most challenging NP-hard problems effectively.

5.4 Chaotic Artificial Bee Colony (CABC) Algorithm

Introduces chaos theory into the ABC algorithm to enhance its exploration capabilities. Chaotic maps can replace or modify certain components of the algorithm, such as the initialization process or the solution update rules.

Jin et al. [27] innovation aims to improve clustering accuracy and efficiency by addressing the conventional K-Means algorithm's limitations, like dependence on initial conditions and susceptibility to local optima. The CAABC algorithm incorporates chaotic maps and an adaptive search strategy to improve global search capabilities and convergence speed. The effectiveness of this approach is demonstrated through comparisons with other heuristic algorithms across standard test functions and datasets, showcasing its advantages in convergence speed and accuracy for solving clustering problems.

Oliva et al. [28] introduced an algorithm designed for the parameter estimation of solar cells, aiming to improve their design and efficiency. The approach combines chaotic maps with the Artificial Bee Colony method, offering a robust solution to optimize the complex and multimodal objective functions involved in solar cell modeling. The CIABC algorithm enhances the search for global solutions, even in the presence of noise, making it a promising tool for designing more efficient photovoltaic modules.

Bharti and Singh [29] present an enhanced version of the Artificial Bee Colony (ABC) algorithm, incorporating chaotic maps and gradient search methods to improve text clustering. The modified algorithm, named Chaotic Gradient Artificial Bee Colony (CGABC), aims to balance exploration and exploitation in search space more effectively. Through empirical evaluations on benchmark text datasets, the CGABC algorithm demonstrates superior clustering performance in terms of solution quality and convergence speed, compared to standard ABC and other variations. The integration of chaotic local search and gradient search methods into the ABC framework significantly enhances its exploitation capabilities, making CGABC a promising approach for text clustering tasks.

5.5 Hybrid Artificial Bee Colony Algorithms

These variants combine the ABC algorithm with other optimization techniques or heuristics, such as genetic algorithms, differential evolution, or local search strategies, to leverage the strengths of multiple approaches.

Jadon et al. [11] introduce a hybrid of ABC and Differential Evolution (DE) algorithms to address the issues of premature convergence, unbalanced exploration–exploitation, and sometimes slow convergence speed in both algorithms. The hybrid algorithm, named HABCDE, modifies the onlooker bee phase of ABC with inspiration from DE and incorporates the best individual concept in the employed bee phase and a modified scout bee phase for higher exploration. Tested over 20 problems and 4 real-world optimization problems, HABCDE shows superior performance compared to basic ABC and DE, as well as other state-of-the-art algorithms like CMA-ES, PSO, BBO, and SMO. The results indicate HABCDE's potential as a competitive meta-heuristic algorithm in the field of optimization.

Ma et al. [12] research presents a hybrid artificial bee colony (HABC) optimizer that dynamically varies population size and ensures a balance between exploration and exploitation by incorporating a bee life-cycle model and comprehensive learning. Powell's search is used to exploit promising areas deeply, while the global best solution information is integrated into the search equation for exploration. HABC demonstrated superior performance over other bio-mimetic algorithms in CEC 2014 benchmarks, showcasing the effectiveness of the proposed hybridization scheme.

Sharma and Pant [30] in their study, introduce a hybrid of the ABC and shuffled frog-leaping algorithm (SFLA), aiming to increase global convergence and balance exploration and exploitation by partitioning the population into superior and inferior groups. The hybrid algorithm, Shuffled-ABC, applies ABC to the superior group and SFLA to the inferior group. Tested on CEC 2006 benchmarks and chemical engineering problems, Shuffled-ABC demonstrates its efficacy in handling constraints and shows promising results compared to classical variants of ABC.

These works contribute significantly to the optimization field by enhancing the Artificial Bee Colony algorithm through hybridization with other meta-heuristic algorithms, demonstrating improved convergence speeds, solution quality, and a better balance between exploration and exploitation capabilities.

5.6 Adaptive Artificial Bee Colony (AABC) Algorithm

This variant introduces adaptive mechanisms to dynamically adjust algorithm parameters, such as the probability of selecting a food source or the limit for abandoning a food source, based on the performance of the search process. The study by Yurtkuran and Emel [31] introduces an adaptive AABC algorithm that utilizes six distinct search strategies for searching and sharing information. A probabilistic selection applied to determine the search rule for generating a candidate solution is updated based on prior performance using the roulette wheel technique. Experiments on benchmark problems with varying dimensionality show that AABC outperforms other ABC variants, demonstrating its effectiveness in solving complex optimization.

Xue et al. [32] introduce a self-adaptive ABC algorithm (SABC-GB) focusing on global optimization. The innovation lies in a solution search equation that utilizes global best candidates and an adaptive mechanism adjusting search strategies based on prior performance. Tested on 25 benchmark functions and applied to a real clustering problem, SABC-GB demonstrates superiority over other algorithms, highlighting its effectiveness for complex optimization problems.

Durgut and Aydin [33] propose an adaptive approach for the ABC algorithm to enhance efficiency in solving binary problems. By incorporating multiple and complementary binary operators and employing three prominent operator selection schemes, the approach achieves superior performance in incapacitated facility location problems, a well-known NP-Hard combinatorial problem. The adaptive hybrid ABC demonstrates its capability to outperform state-of-art algorithms, showcasing its applicability and efficiency in binary optimization tasks.

5.7 Discrete Artificial Bee Colony (DABC) Algorithm

DABC algorithm is specifically designed for discrete optimization problems, this variant adapts the solution representation and update mechanisms to suit discrete search spaces.

Yurtkuran and Emel [34] present a discrete ABC algorithm to address single-machine earliness-tardiness scheduling problems. By employing multiple search operators and introducing novel crossover operators, the algorithm achieves a balanced exploration and exploitation behavior. The performance is evaluated on benchmark problems and compared with state-of-the-art algorithms, showing the proposed meta-heuristic's ability to produce better solutions in terms of quality, robustness, and computational time, demonstrating its potential for efficiently solving combinatorial optimization problems.

Zhang and Liu [35] integrate a discrete search equation to develop a discrete improved artificial bee colony (DiABC) algorithm aimed at optimizing the 0–1 knapsack problem. Enhancements include initializing with a random binary population, introducing the current optimal solution into search equations for worker and onlooker bees, and proposing a new selection method for scout bees. Tested across 18 benchmark knapsack problems, DiABC validated its effectiveness, outperforming traditional approaches in optimizing complex combinatorial tasks.

These variants demonstrate the flexibility of the ABC algorithm and its adaptability to a wide range of optimization problems. Researchers continue to develop new variants and improvements to further enhance its performance and applicability (Table 1).

Table 1 Description of variants of the ABC algorithm

Serial number	Reference	Variants	Key findings	Application
1	[22]	GABC	Incorporates "Fully Informed Learning" in the onlooker bee phase to mitigate local optima trapping. Shows competitive performance	Complex optimization problems
2	[21]	GABC	Adjusts step size iteratively for constrained optimization. Demonstrates superior performance	Constrained and unconstrained optimization problems
3	[20]	GABC	Improves exploitation capability for constrained optimization. Enhancements introduce diversification and intensification	Constrained optimization problems
4	[24]	MABC	Incorporates directional information for improved convergence. Enhanced solution quality and robustness	Global or near-global optimization
5	[23]	MABC	Uses chaotic mapping and global solution introduction to escape local optima. Demonstrated higher accuracy	Binary optimization problems
6	[15]	BABC	IDisABC tackles the challenge of limited solution generation through genetically inspired components. Demonstrates superiority in dynamic clustering	Dynamic clustering, handling one of the most challenging NP-hard problems effectively
7	[25]	BABC	NBABC introduces a dimension limitation mechanism. Demonstrates competitive performance, outperforming other algorithms in certain cases	Binary optimization in general (e.g., OneMax problem, 0/1 Knapsack problems, Feature Selection)
8	[26]	BABC	IbinABC enhances exploration to avoid local minima with a fitness value-based update mechanism. Shows superiority in convergence speed and robustness	Binary optimization problems, specifically the incapacitated facility location problem
9	[29]	CABC	Incorporates chaotic maps and gradient search methods to balance exploration and exploitation effectively. Shows superior clustering performance in terms of solution quality and convergence speed	Text clustering with enhanced solution quality and convergence speed
10	[28]	CABC	Combines chaotic maps with ABC for parameter estimation of solar cells, enhancing the search for global solutions even in noisy environments. Promises more efficient photovoltaic module design	Parameter estimation for the design and efficiency improvement of solar cells

(continued)

Table 1 (continued)

Serial number	Reference	Variants	Key findings	Application
11	[27]	CABC	Improves clustering accuracy and efficiency by incorporating chaotic maps and adaptive search strategy. Demonstrates advantages in convergence speed and accuracy for solving clustering problems	Improving K-Means clustering accuracy and efficiency
12	[12]	HABC	HABC dynamically varies population size and incorporates a bee life-cycle model for balance between exploration and exploitation. Demonstrated superior performance in benchmarks	CEC 2014 benchmarks for bio-mimetic algorithms
13	[11]	HABC	HABCDE addresses premature convergence and unbalanced exploration–exploitation. Shows superior performance over ABC, DE, and other algorithms	General optimization problems
14	[30]	HABC	Shuffled ABC increases global convergence and balances exploration and exploitation by partitioning the population. Shows efficacy in handling constraints	CEC 2006 benchmarks and chemical engineering problems
15	[31]	AABC	Employs six different search rules with probabilistic selection based on performance. Outperforms other ABC variants in global optimization	Complex optimization problems
16	[32]	AABC	Utilizes global best candidates with an adaptive mechanism for search strategy adjustments. Demonstrates superiority in complex optimization	Global optimization, including real clustering problems
17	[33]	AABC	Incorporates multiple binary operators and operator selection schemes for binary optimization. Achieves superior performance in combinatorial problems	Binary optimization in combinatorial problems
18	[34]	DABC	Employs multiple search operators and introduces novel crossover operators for balanced exploration and exploitation. Demonstrates potential for efficiently solving combinatorial optimization problems	Single machine earliness-tardiness scheduling problems
19	[35]	DABC	Integrates a discrete search equation with a random binary population and a new selection method for scout bees. Outperforms traditional approaches in optimizing complex combinatorial tasks	0–1 Knapsack problem

6 Conclusion

The Artificial Bee Colony (ABC) algorithm, renowned for its optimization capability, has evolved through various enhancements to address diverse optimization challenges. These improvements focus on enhancing exploration and exploitation abilities, tailoring solutions to specific domains, and striking a balance between these aspects. In this chapter, we have covered the end-to-end working mechanism of the ABC algorithm and its seven prominent variants proposed to cater specific applications to diverse areas. The GABC algorithm introduces a global best solution concept, refining search processes and bolstering optimal solution exploitation. MABC algorithms, incorporating features like chaotic mapping and directional information, augment exploration and solution. BABC algorithms, catering to binary optimization, employ innovative mechanisms for robust solution generation. CABC algorithms infuse chaos theory, enhancing exploration and convergence. Hybrid ABC algorithms amalgamate ABC with other heuristics, yielding superior convergence and exploration. Adaptive ABC algorithms dynamically adjust parameters to optimize performance. Discrete ABC algorithms tailor solutions for solving combinatorial optimization problems, showcasing remarkable performance in discrete search space. These variants underscore the ABC algorithm's adaptability and efficiency in solving a myriad of optimization challenges, continuing to inspire novel advancements in the field.

References

1. Karaboga D (2005) An idea based on honey bee swarm for numerical optimization. Technical report-tr06, Erciyes university, engineering faculty, computer
2. Singh A (2009) An artificial bee colony algorithm for the leaf-constrained minimum spanning tree problem. Appl Soft Comput 9(2):625–631
3. Karaboga D, Akay B (2009) A comparative study of artificial bee colony algorithm. Appl Math Comput 214(1). https://doi.org/10.1016/j.amc.2009.03.090
4. Karaboga D, Basturk B (2007b) Artificial bee colony (ABC) optimization algorithm for solving constrained optimization problems. Lecture notes in computer science (Including subseries lecture notes in artificial intelligence and lecture notes in bioinformatics). LNAI, vol 4529. https://doi.org/10.1007/978-3-540-72950-1_77
5. Akay B, Karaboga D (2012) Artificial bee colony algorithm for large-scale problems and engineering design optimization. J Intell Manuf 23:1001–1014
6. Karaboga D, Gorkemli B, Ozturk C, Karaboga N (2014) A comprehensive survey: artificial bee colony (ABC) algorithm and applications. Artif Intell Rev 42(1). https://doi.org/10.1007/s10462-012-9328-0
7. Hegazy O, Soliman OS, Salam MA (2014) LSSVM-ABC algorithm for stock price prediction. arXiv:1402.6366
8. Hegazy O, Soliman OS, Salam MA (2015) Comparative study between FPA, BA, MCS, ABC, and PSO algorithms in training and optimizing of LS-SVM for stock market prediction. Int J Adv Comput Res 5(18):35–45
9. Liu J-X, Jia Z-H (2018) Telecommunication traffic prediction based on improved LSSVM. Int J Pattern Recognit Artif Intell 32(03):1850007

10. Paliwal P, Kumar D (2018) ABC based neural network approach for churn prediction in telecommunication sector. Inf Commun Technol Intell Syst (ICTIS 2017) 2(2):343–349
11. Jadon SS, Tiwari R, Sharma H, Bansal JC (2017) Hybrid artificial bee colony algorithm with differential evolution. Appl Soft Comput J 58. https://doi.org/10.1016/j.asoc.2017.04.018
12. Ma L, Zhu Y, Zhang D, Niu B (2016) A hybrid approach to artificial bee colony algorithm. Neural Comput Appl 27(2). https://doi.org/10.1007/s00521-015-1851-x
13. Karaboga D, Basturk B (2007a) A powerful and efficient algorithm for numerical function optimization: artificial bee colony (ABC) algorithm. J Glob Optim 39(3). https://doi.org/10.1007/s10898-007-9149-x
14. Karaboga D, Basturk B (2008) On the performance of artificial bee colony (ABC) algorithm. Appl Soft Comput J 8(1). https://doi.org/10.1016/j.asoc.2007.05.007
15. Ozturk C, Hancer E, Karaboga D (2015) Dynamic clustering with improved binary artificial bee colony algorithm. Appl Soft Comput J 28. https://doi.org/10.1016/j.asoc.2014.11.040
16. Csilléry K, Blum MG, Gaggiotti OE, François O (2010) Approximate Bayesian computation (ABC) in practice. Trends Ecol Evol 25(7):410–418
17. Karaboga D, Akay B (2011) A modified artificial bee colony (ABC) algorithm for constrained optimization problems. Appl Soft Comput 11(3):3021–3031
18. Hancer E, Xue B, Karaboga D, Zhang M (2015) A binary ABC algorithm based on advanced similarity scheme for feature selection. Appl Soft Comput 36:334–348
19. Aslan S (2020) A comparative study between artificial bee colony (ABC) algorithm and its variants on big data optimization. Memetic Comput 12(2):129–150
20. Sharma S, Kumar S, Sharma K (2021) Improved Gbest artificial bee colony algorithm for the constraints optimization problems. Evol Intell 14(3). https://doi.org/10.1007/s12065-019-00231-8
21. Bhambu P, Sharma S, Kumar S (2018) Modified Gbest artificial bee colony algorithm. Adv Intell Syst Comput 583. https://doi.org/10.1007/978-981-10-5687-1_59
22. Sharma K, Gupta PC, Sharma H (2016) Fully informed artificial bee colony algorithm. J Exp Theor Artif Intell 28(1–2). https://doi.org/10.1080/0952813X.2015.1056238
23. Wang Y, You J, Hang J, Li C, Cheng L (2018) An improved artificial bee colony (ABC) algorithm with advanced search ability. In: Proceedings of 2018 IEEE 8th international conference on electronics information and emergency communication, ICEIEC 2018. https://doi.org/10.1109/ICEIEC.2018.8473513
24. Kiran MS, Findik O (2015) A directed artificial bee colony algorithm. Appl Soft Comput J 26. https://doi.org/10.1016/j.asoc.2014.10.020
25. Santana CJ, Macedo M, Siqueira H, Gokhale A, Bastos-Filho CJA (2019) A novel binary artificial bee colony algorithm. Futur Gener Comput Syst 98. https://doi.org/10.1016/j.future.2019.03.032
26. Durgut R (2021) Improved binary artificial bee colony algorithm. Front Inf Technol Electron Eng 22(8). https://doi.org/10.1631/FITEE.2000239
27. Jin Q, Lin N, Zhang Y (2021) K-means clustering algorithm based on chaotic adaptive artificial bee colony. Algorithms 14(2). https://doi.org/10.3390/a14020053
28. Oliva D, Ewees AA, Aziz MAE, Hassanien AE, Cisneros MP (2017) A chaotic improved artificial bee colony for parameter estimation of photovoltaic cells. Energies 10(7). https://doi.org/10.3390/en10070865
29. Bharti KK, Singh PK (2016) Chaotic gradient artificial bee colony for text clustering. Soft Comput 20(3). https://doi.org/10.1007/s00500-014-1571-7
30. Sharma TK, Pant M (2017) Shuffled artificial bee colony algorithm. Soft Comput 21(20). https://doi.org/10.1007/s00500-016-2166-2
31. Yurtkuran A, Emel E (2015) An adaptive artificial bee colony algorithm for global optimization. Appl Math Comput 271. https://doi.org/10.1016/j.amc.2015.09.064
32. Xue Y, Jiang J, Zhao B, Ma T (2018) A self-adaptive artificial bee colony algorithm based on global best for global optimization. Soft Comput 22(9). https://doi.org/10.1007/s00500-017-2547-1

33. Durgut R, Aydin ME (2021) Adaptive binary artificial bee colony algorithm. Appl Soft Comput 101. https://doi.org/10.1016/j.asoc.2020.107054
34. Yurtkuran A, Emel E (2016) A discrete artificial bee colony algorithm for single machine scheduling problems. Int J Prod Res 54(22). https://doi.org/10.1080/00207543.2016.1185550
35. Zhang S, Liu S (2019) A discrete improved artificial bee colony algorithm for 0-1 knapsack problem. IEEE Access 7. https://doi.org/10.1109/ACCESS.2019.2930638
36. Comert SE, Yazgan HR (2023) A new approach based on hybrid ant colony optimization-artificial bee colony algorithm for multi-objective electric vehicle routing problems. Eng Appl Artif Intell 123(Part B): 106375, ISSN 0952-1976. https://doi.org/10.1016/j.engappai.2023.106375
37. Wang J, Liu Y, Rao S, Zhou X, Hu J (2023) A novel self-adaptive multi-strategy artificial bee colony algorithm for coverage optimization in wireless sensor networks. Ad Hoc Networks 150: 103284, ISSN 1570-8705, https://doi.org/10.1016/j.adhoc.2023.103284
38. Chen L, Wu T, Wang Z, Lin X, Cai Y (2023) A novel hybrid BPNN model based on adaptive evolutionary Artificial Bee Colony Algorithm for water quality index prediction. Ecolog Indicators 146: 109882, ISSN 1470-160X, https://doi.org/10.1016/j.ecolind.2023.109882

The Rise of Artificial Bee Colony Algorithms in Data Science and Machine Learning is Notable

Arif Md. Sattar, Mritunjay Kr. Ranjan, Shilpi Saxena, Shreya Tiwari, and Sanjay Kr. Tiwari

1 Introduction

The Artificial Bee Colony (ABC) algorithm has been a game-changer in addressing difficult optimization issues, even though bee buzzing sounds seem unrelated to computing. Several fields, including engineering and finance, have discovered uses for this metaheuristic algorithm, which was inspired by honeybee foraging behavior [1]. The ABC algorithm will be dissected in this chapter, including its history, inner workings, and several applications to real-world issues. One of the most talked-about optimization algorithms recently is the ABC model, which is based on honeybees' intricate foraging habits. The ABC algorithm's exceptional capacity to rapidly traverse complicated optimization spaces is particularly noteworthy, given that nature has frequently been the wellspring of novel methods of problem-solving [2]. By delving into its basic concepts, methods, and the wide range of problems it has effectively solved across different industries, this chapter investigates the revolutionary uses of the ABC algorithm. The ABC algorithm is based on the idea that robots and bees work together to solve problems, much like a team. The algorithm mimics the foraging dynamics of honeybee colonies by separating its agents into worker

A. Md. Sattar (✉)
Diwan V.S. Institute of Management Studies, Meerut, India
e-mail: amsattargaya@gmail.com

M. Kr. Ranjan
School of Computer Science and Engineering, Sandip University, Nashik, Maharashtra, India

S. Tiwari
Department of Computer Science and Engineering, Amity School of Engineering and Technology, Amity University, Noida, India

S. Kr. Tiwari
Department of Mathematics, Magadh University, Bodh-Gaya, India

S. Saxena
Department of Computer Application & IT, Lords University, alwar, Rajasthan, India

K. Raza (ed.), *Solving with Bees*, Springer Tracts in Nature-Inspired Computing,
https://doi.org/10.1007/978-981-97-7344-2_2

bees, observer bees, and scout bees. In these colonies, the bees cooperate together to discover the best food sources. To understand the mentality that drives the ABC algorithm to solve complex optimization problems, it is essential to understand this biological inspiration [3]. As we go through the chapters, we'll examine the ABC algorithm's main parts, breaking it down into the responsibilities of the employed bees, who use the information they receive to exploit current solutions, the spectator bees, who choose solutions based on that information, and the scout bees, who explore new possibilities. Engineering design problems, financial forecasting issues, and healthcare optimization are just a few examples of the real-world problems that this investigation lays the groundwork for addressing. The importance of the ABC method in domains where identifying the optimal solution from a large pool of options is critical will be uncovered in the sections that follow as we explore the optimization landscape in which it excels. Additionally, we will go into the difficulties of parameter adjustment, highlighting the continuous endeavors to enhance the algorithm for maximum efficiency. The goal of this chapter is to give a thorough grasp of how the ABC algorithm simplifies complex problems, making it a leader in optimization algorithms [4], by guiding the reader through case studies and real-world applications.

As we go through the chapters, we'll examine the ABC algorithm's main parts, breaking it down into the responsibilities of the employed bees, who use the information they receive to exploit current solutions, the spectator bees, who choose solutions based on that information, and the scout bees, who explore new possibilities [5]. Engineering design problems, financial forecasting issues, and healthcare optimization are just a few examples of the real-world problems that this investigation lays the groundwork for addressing. The importance of the ABC method in domains where identifying the optimal solution from a large pool of options is critical will be uncovered in the sections that follow as we explore the optimization landscape in which it excels. Additionally, we will go into the difficulties of parameter adjustment, highlighting the continuous endeavors to enhance the algorithm for maximum efficiency [6]. Through the exploration of real-world examples and case studies, this chapter seeks to offer a thorough grasp of how the ABC algorithm simplifies complex problems, establishing it as a leader among optimization algorithms. The objectives of ABC algorithms are as follows:

a. To efficiently explore complex search spaces using honeybee foraging behavior to provide near-optimal solutions in engineering, data mining, and AI.
b. To improve machine learning model performance and accuracy by addressing feature selection, parameter optimization, model selection, and ensemble learning optimization challenges.
c. Navigating complex, multidimensional landscapes in machine learning and other fields because it balances exploration and exploitation.

Table 1 Comparison with respect to methods: pros, cons, and research gaps

Methods	Pros	Cons	Research gap	References
Random forest, logistic regression, ABC algorithm	Integrates multiple algorithms for hepatitis C virus detection, potentially improving accuracy	Potential complexity in model interpretation due to the combination of algorithms	Further validation and comparison with existing models on larger datasets needed	[7]
Parallel artificial bee colony algorithm	Offers PAPR reduction in OFDM and MIMO-OFDM systems, enhancing signal quality	May require significant computational resources for parallelization	Investigation into real-world implementation and scalability required	[8]
Hybrid ABC-DE algorithm	Combines the strengths of ABC and DE algorithms for improved pattern synthesis	Combines the strengths of ABC and DE algorithms for improved pattern synthesis	Evaluation of algorithm performance across different array configurations needed	[9]
Agent-based ABC algorithm	Enables parallel processing for hyperspectral image endmember extraction, speeding up computation	Potential scalability challenges in large-scale hyperspectral datasets	Exploration of agent-based ABC algorithm efficiency in handling diverse image characteristics	[10]
K-means clustering algorithm, improved ABC algorithm	Offers improved clustering of wind power scenarios, aiding in resource planning and management	Sensitivity to initialization parameters and cluster shape variations	Comparative study with other clustering algorithms to assess performance and robustness	[11]

2 Related Work

It helps to find holes in the body of current research, give their own studies a critical context, and show that they know about the relevant theories and methods. It helps researchers situate their work within the broader scholarly conversation, demonstrate their understanding, and refine their research objectives in Table 1.

3 Action Implied in ABC Algorithm

Exploitation and exploration are both important parts of the ABC algorithm, which makes it fast and easy to find the solution space for the best answers. Exploitation uses information from good solutions to make them even better, while discovery adds variety by coming up with new possible solutions [12]. The ABC algorithm successfully searches the space for solutions and gets closer to good ones by changing the balance between these two acts on the fly.

3.1 Foraging Behavior of Bees

Honeybee foraging behavior is the natural inspiration for the ABC algorithm, which may be understood by studying it. Bees efficiently locate the best food sources in their habitat through a combination of group communication and investigation [13]. The design of the ABC algorithm is based on this collaborative and decentralized approach. An optimization technique that draws inspiration from nature, the ABC algorithm is based on the way bees forage. A metaheuristic algorithm that can traverse complex solution spaces has been developed based on the complex and efficient foraging techniques used by honeybees. To fully grasp the inner workings and dynamics of the ABC algorithm, one must have a firm grasp of bee foraging behavior in Table 2.

3.2 Core Components of ABC

The algorithm consists of three main components: employed bees, onlooker bees, and scout bees. Employed bees exploit current solutions, onlooker bees select solutions based on the employed bees' information, and scout bees explore new solutions. These interactions simulate the foraging dynamics seen in bee colonies. The ABC algorithm derives its effectiveness from its ingenious simulation of the foraging behavior of honeybee colonies [19]. By dissecting its core components, we gain insight into the intricate mechanisms that enable the algorithm to navigate complex optimization spaces with efficiency and adaptability (Table 3).

4 ABC Algorithm Optimization Landscape in Action

"ABC Algorithm in Action" shows how the ABC algorithm is used to solve optimization problems in the real world. It does this by using the intelligence of robotic bees to quickly search solution spaces and find optimal or nearly optimal solutions. The optimum answer from a huge number of options is what optimization problems are all about, and the ABC algorithm is very good at doing that. In many areas, like engineering design, transportation planning, and resource sharing, it can be used. The ABC algorithm is a nature-inspired optimization method that is based on how honeybees find food. It's used a lot to solve different kinds of planning tasks [24]. The surface made by the problem's goal function is called the optimization landscape in ABC. The objective function shows how good or fit an answer is. The goal of the

Table 2 Attribute and impact

Attribute	Impact of attribute	References
Collective intelligence	Bee colonies are strong because of their intelligence. Bees collaborate and communicate to make decisions. In the ABC algorithm, artificial bees collaborate to find the best solution to a problem	[14]
Exploration and exploitation	Bee colonies balance exploration and exploitation. While employed bees use known food sources, scout bees search for new ones. The ABC algorithm uses employed bees to exploit current solutions and scout bees to discover new ones. To navigate the optimization landscape, the algorithm dynamically balances exploration and exploitation	[15]
Dance communication	Complex dances help honeybees find food. This communication system informs the colony of promising food sources. According to Tong, the ABC algorithm simulates information sharing by bees communicating solution quality to other bees. Through indirect communication, optimal solutions can be found collaboratively	[16]
Adaptability	Bee colonies adapt to environmental changes and food shortages. The ABC algorithm dynamically adjusts parameters like scout bees and exploration rate to optimize its search strategy. This adaptability lets the algorithm adapt to changing optimization problem characteristics	[17]
Decentralized decision-making	Bee colonies decide decentralized. No one controls foraging; each bee makes decisions based on local information. Artificial bees act independently in the ABC algorithm, reflecting its decentralization. This decentralized approach improves algorithm scalability and robustness. Bee foraging behavior illuminates the ABC algorithm's design and operation. The algorithm uses collective intelligence, exploration–exploitation balance, communication mechanisms, adaptability, and decentralized decision-making to solve complex optimization problems across domains. Bee foraging behavior, elegantly translated into algorithmic processes, shows how bio-inspired approaches can solve real-world problems	[18]

Table 3 Properties and explanations

Properties	Explanation	References
Employed bees: the workforce of optimization	The ABC algorithm relies on employed bees, the workforce that exploits current solutions. These worker bees forage for food in a bee colony. The algorithm uses bees to carry optimization solutions and iteratively explore the neighborhood for better solutions. Exploitation involves assessing these solutions, comparing them, and updating positions based on performance. The bees enable iterative improvement, ensuring the algorithm converges to optimal solutions	[20]
Onlooker bees: selective decision-makers	The ABC algorithm's collective decision-making relies on observer bees. They represent the bees observing the employed bees and making decisions based on the data. Onlooker bees choose solutions from employed bees' repertoire, emphasizing collaboration and decentralization. Performance affects selection, with better solutions being chosen more often. Onlooker bees guide the algorithm to promising solution space regions, balancing exploration and exploitation	[21]
Scout bees: exploration pioneers	The ABC algorithm uses scout bees, like exploratory bees in a natural colony, to diversify solution search. Scout bees experiment with new solutions to add randomness and innovation. Scout bees look for new solutions when a solution fails to improve after a certain number of iterations. This exploration mechanism prevents premature convergence and helps the algorithm escape local optima, improving global optimization	[22]
Collective dynamics and communication	The dynamic interactions between employed, onlooker, and scout bees make the ABC algorithm successful. This decentralized approach lets the algorithm harness collective intelligence, where information flows seamlessly through the population. Employed bees inform onlooker bees of their findings to aid selection. Scout bees explore new solutions to bring diversity and balance exploration and exploitation	[23]

algorithm in the case of ABC is to look through this landscape to find the best answer that either minimizes or maximizes the target function. The ABC algorithm works with a group of fake bees, and each bee is a possible answer to the optimization problem. These answers are known as food sources. There are three major stages in the algorithm: the hired bee phase, the onlooker bee phase, and the scout bee phase in Table 4.

The optimization environment is shaped by the ABC algorithm's intrinsic exploration–exploitation trade-off. Utilizing advantageous areas, employed and observer

Table 4 Phase action and algorithmic approach

Phase	Action	Algorithmic approach	References
Employed bee phase	For each employed bee	S_1. Explore the neighborhood of the current solution by adjusting its position S_2. Evaluate the fitness of the solution using the objective function S_3. If the new solution is better than the current one: a. Update the bee's position b. Otherwise, retain the current state	[25]
Onlooker bee phase		S_1. Onlooker bees select food sources based on information from employed bees S_2. Calculate the probability of choosing each food source based on its fitness S_3. Select a food source probabilistically; better solutions have higher probabilities S_4. Explore the neighborhood of the selected solutions, like employed bees	[26]
Scout bee phase		If an employed bee exhausts its search and cannot improve its solution: S_1. Bee transitions into a scout bee Scout Bee: S_2. Randomly explore new solutions S_3. Introduce diversity into the population	[27]

bees investigate the terrain, whereas scout bees initiate exploration by arbitrarily seeking out novel solutions [28]. The effectiveness of the ABC algorithm is contingent on its capacity to efficiently balance exploration and exploitation. With the objective of efficiently traversing the optimization landscape, the algorithm progressively approaches the global optimum while evading local minima. Scholars frequently investigate the convergence characteristics, the deal between exploration and exploitation, and the influence of different parameters on the optimization domain in order to improve the algorithm's performance for particular problem domains. The evaluation of the optimization landscape offers valuable insights regarding the ABC algorithm's capability to manage various types of objective functions and facilitates the refinement of its parameters to achieve optimal outcomes.

5 Parameter Tuning

One key challenge in applying the ABC algorithm lies in tuning its scope to achieve optimal performance. Researchers have dug out various approaches, such as adaptive parameter adjustment and hybridization with other algorithms, to enhance the algorithm's efficiency and effectiveness. In order to optimize the efficacy of any algorithm, scope tuning is essential [29]. This principle acts in the ABC algorithm as well. ABC, which finds inspiration in the foraging behavior of honeybees, is an extensively employed method for resolving a multitude of optimization problems. Properly tuning its parameters can significantly impact its convergence speed and solution quality.

Comprehensive breakdown of the ABC algorithm's critical parameters:

a. ***Number of Employed Bees (N_e)***

This parameter represents the total number of employed bees in the colony. Increasing N_e may enhance exploration capability but it also leads to a higher computational cost. Obtaining the supremum value is frequently contingent on the problem's complexity and the computational resources at hand.

$$N_e = \frac{N_{total}}{2} \tag{1}$$

where N_{total}, represents the total number of bees in the colony, including both employed and onlooker bees.

b. ***Number of Onlooker Bees (N_o)***

Onlooker bees are responsible for choosing food sources based on the waggle dance information. Adjusting N_o influences the exploitation-exploitation balance. A higher value may intensify exploitation, while a lower value may emphasize exploration. It is typically set proportional to N_e.

$$N_o = N_e \tag{2}$$

where, N_e represents the number of onlooker bees.

c. ***Number of Scout Bees (N_S)***

Scout bees are tasked with discovering new food sources. The parameter N_s determines the number of employed bees whose solutions are abandoned. A higher N_s encourages more exploration but might slow down convergence. It is often set as a fixed percentage of N_e.

$$N_s = P * N_e \tag{3}$$

where P represents the fixed percentage of employed bees whose solutions are abandoned by scout bees. And N_e represents the number of employed bees.

d. ***Limit (L)***

L is the limit parameter, representing the number of iterations for which a food source is allowed to persist without improvement. After L iterations without improvement, a scout bee is sent to discover a new solution. Properly setting L is essential to balance exploration and exploitation.

$$L = Constant_1 \tag{4}$$

where L represents the limit parameter, which is the number of iterations for which a food source is allowed to persist without improvement. The value of L is typically set as a constant determined by the user or problem requirements.

e. ***Trial Limit (T)***

T is the maximum number of trials for each employed or onlooker bee to find a better solution. A higher T can enhance exploration but may increase computational overhead. The optimal value depends on the problem characteristics and computational resources.

$$T = Constant_2 \tag{5}$$

where T represents the trial limit parameter, which is the maximum number of trials for each employed or onlooker bee to find a better solution. The value of T is typically set as a constant determined by the user or problem requirements.

f. ***Initialization Range***

The initial range of solutions greatly influences the search for space exploration. Adjusting the range should be problem-specific. A narrow range might lead to premature convergence, while a broad range might hinder convergence.

$$Initilaization\ Range = [a, b] \tag{6}$$

$[a, b]$ represents the range within which the initial solutions are randomly generated. Where a and b are the lower and upper bounds of the initialization range, respectively.

g. ***Food Source Update Strategy***

The way in which the employed bees modify the solutions and the onlooker bees choose food sources can be fine-tuned. For instance, exploring different mutation or crossover strategies and determining the probability of selecting a food source based on fitness.

$$Food\ Source\ Update\ Strategy = Strategy\ (Parameters) \tag{7}$$

where Strategy represents the specific update strategy used by the employed bees to modify solutions and by the onlooker bees to choose food sources. Parameters denote the parameters or configurations associated with the chosen update strategy, such as mutation rates, crossover strategies, or selection probabilities based on fitness.

h. ***Fitness Function Scaling***

Scaling the fitness function is essential to ensure that the algorithm converges effectively. Normalizing the fitness values helps in achieving a balanced exploration-exploitation deal off.

$$\textit{Selected Fitness}\,(f) = \frac{f - \min}{\max - \min} \tag{8}$$

where f represents the original fitness value obtained from the fitness function evaluation. min and max represent the minimum and maximum fitness values observed in the population, respectively. Scaled Fitness (f) represents the scaled fitness value, which is normalized to fall within the range of [0, 1].

When tuning parameters for the ABC algorithm, it is often recommended to conduct multiple experiments with different parameter configurations and evaluate their performance on benchmark problems. This iterative process helps in identifying the optimal set of parameters for a specific optimization task.

6 Applications Across Industries

A mechanism of optimization inspired by the foraging behavior of honeybees; the ABC algorithm was developed. The algorithm has been widely implemented in numerous industries as a result of its effective resolution of intricate optimization challenges. Consider the following implementations in various industries (Table 5).

The versatility of the ABC algorithm makes it a valuable tool in addressing optimization challenges across a wide range of industries, contributing to enhanced efficiency and resource utilization.

Table 5 Field, focused optimization, and description

Field	Focused	Description
Engineering and design [30]	Structural optimization	ABC algorithm is used in optimizing the design of structures, such as bridges and buildings, by minimizing material usage while ensuring structural integrity
	Mechanical design	It aids in optimizing the parameters of mechanical systems, like gear ratios or component dimensions, to enhance performance and efficiency
Communication networks [31]	Routing optimization	In telecommunication networks, ABC can be used to optimize routing paths, minimizing delays and packet loss
	Wireless sensor networks	ABC wireless sensor networks: ABC helps in optimizing the imposed of sensors to maximize coverage and reduce energy consumption in wireless sensor networks
Finance power systems [32]	Portfolio optimization	ABC acts in financial portfolio management to optimize investment portfolios by considering risk and return factors
	Risk management	It assists in risk assessment and management by optimizing risk models and predicting potential financial risks
	Load dispatch optimization	ABC is utilized to optimize the distribution of power loads in electrical grids, leading to improved energy efficiency
	Renewable energy	In the planning of renewable energy systems, ABC helps in optimizing the placement and capacity of renewable energy sources
Medical and healthcare [33]	Treatment planning	ABC is used for maximizing treatment plans in medical fields, such as radiation therapy, to achieve optimal dosages while minimizing side effects
	Drug discovery	In pharmaceutical research, ABC aids in optimizing the parameters of drug formulations and predicting molecular structures
Manufacturing and supply chain [34]	Production scheduling	ABC optimizes production schedules in manufacturing, considering factors like machine utilization and production costs
	Supply chain optimization	It helps in maximizing supply chain logistics, including inventory management, transportation routes, and order fulfillment

(continued)

Table 5 (continued)

Field	Focused	Description
Data mining and machine learning [35]	Feature selection	ABC is applied in selecting the most relevant features in datasets, improving the efficiency of machine learning models
	Hyperparameter tuning	It is used for optimizing hyperparameters of machine learning algorithms, leading to better model performance
Robotics [36]	Path planning	ABC aids in optimizing the path planning of robots, ensuring efficient and collision-free navigation in various environments

7 Challenges of ABC Algorithm

Let's delve into the challenges faced by the ABC algorithm. While ABC is a popular optimization algorithm inspired by the foraging behavior of honeybees, it is not without its set of challenges (Table 6).

Despite these challenges, ongoing research and advancements in metaheuristic algorithms continue to address and mitigate these issues, making ABC a promising approach for optimization in various domains.

8 Future Directions

The prospective avenues of the ABC algorithm exhibit considerable potential for additional progress in the domains of optimization and resolution of problems. The following are prospective directions for further advancement in Table 7.

As researchers continue to explore these directions, the ABC algorithm is likely to evolve into a more versatile and powerful optimization tool with broader applicability across diverse domains.

Table 6 Challenges of ABC algorithm with description

Challenges	Description
Exploration–exploitation balance [37]	The algorithm must strike a delicate balance between exploration and exploitation. Too much exploration can lead to inefficient searching, while excessive exploitation might cause the algorithm to get stuck in local optima
Parameter tuning [38]	ABC involves several parameters, such as the number of employed bees, onlooker bees, scout bees, and the limit on abandoned solutions. Finding the optimal values for these parameters can be challenging and might require extensive tuning
Dynamic optimization problems [39]	ABC may struggle with dynamic optimization problems where the objective function or constraints change over time. Adapting the algorithm to handle such dynamic environments poses a significant challenge
Convergence speed [40]	ABC may exhibit slower convergence rates compared to some other optimization algorithms. Improving the convergence speed without sacrificing solution quality is an ongoing challenge
Premature convergence [41]	There is a risk of premature convergence, where the algorithm converges to a suboptimal solution before thoroughly exploring the search space. This can limit the algorithm's ability to discover the global optimum
Handling high-dimensional spaces [42]	In high-dimensional optimization problems, ABC may struggle due to the curse of dimensionality. The algorithm might require more iterations to explore the vast search space adequately
Adaptability to problem characteristics [43]	The performance of ABC can vary based on the characteristics of the optimization problem. Designing a version of ABC that adapts well to different types of problems is a challenge
Memory usage	For large-scale optimization problems, the memory requirements of the algorithm can be substantial. Efficient memory management strategies need to be developed to handle such scenarios
Handling constraints [44]	Problems with constraints pose a challenge for ABC. Ensuring the algorithm can handle and satisfy constraints while optimizing the objective function is an ongoing area of research
Robustness [45]	ABC's robustness in noisy or uncertain environments needs improvement. Enhancing its ability to deal with stochasticity and noisy objective functions is crucial for real-world applications

Table 7 The future direction of ABC algorithms with their description

Future attribute	Description
Hybridization with other algorithms [46]	Researchers can explore combining ABC with other nature-inspired algorithms or traditional optimization methods to create hybrid approaches. This can lead to improved performance and robustness across a wider range of problems
Adaptive and dynamic strategies [47]	Developing adaptive and dynamic strategies within the ABC algorithm can enhance its ability to adapt to changing problem landscapes. This could involve dynamically adjusting parameters during runtime based on the characteristics of the optimization problem being solved
Handling constraints and multi-objective optimization [48]	Integrating mechanisms to handle constraints and extending the algorithm to address multi-objective optimization problems are crucial areas for improvement. This would make ABC more versatile in solving real-world problems with complex constraints and multiple conflicting objectives
Parallel and distributed computing [49]	Implementing parallel and distributed versions of ABC can leverage the power of modern computing architectures. This approach can significantly enhance the algorithm's scalability and enable it to tackle larger and more complex optimization tasks
Cognitive-inspired modifications [50]	Exploring modifications inspired by cognitive processes, such as memory and learning mechanisms, can potentially improve the exploration and exploitation balance in ABC. Incorporating these elements may lead to more intelligent and adaptive search behavior
Fine-tuning algorithm parameters [51]	Research can focus on developing automated methods for fine-tuning ABC parameters. This would help practitioners apply the algorithm more effectively to diverse problem domains without the need for extensive manual parameter tuning
Real-world applications [52]	Increasing the focus on applying ABC to real-world applications in various fields, such as engineering, finance, and healthcare, will contribute to validating its effectiveness and identifying areas for further improvement

(continued)

Table 7 (continued)

Future attribute	Description
Benchmarking and standardization [53]	Establishing benchmark problems and standardizing evaluation metrics for ABC will facilitate fair comparisons between different algorithmic variations. This is essential for accurately assessing the algorithm's performance and guiding further development
Explainability and interpretability [54]	Enhancing the interpretability of the ABC algorithm can improve its acceptance and adoption in practical applications. Understanding how the algorithm makes decisions can be crucial in domains where interpretability is a key consideration
Continuous improvement through community collaboration [55]	Encouraging collaboration within the research community and open-sourcing implementations can lead to a collective effort in refining and advancing the ABC algorithm. Shared insights, datasets, and codebases can accelerate progress in this field

9 Conclusion

The Artificial Bee Colony (ABC) algorithm emerges as a transformative tool with immense potential for solving complex optimization problems across various domains. The unique inspiration drawn from the collective foraging behavior of honeybees has proven to be a fruitful foundation for developing efficient optimization techniques. The success of ABC lies in its ability to strike a delicate balance between exploration and exploitation, mimicking the natural foraging behaviors observed in real bee colonies. This adaptability makes it a versatile and robust algorithm, capable of addressing a wide range of optimization challenges. As we delve into the transformative applications of the ABC algorithm, it becomes evident that its impact extends far beyond theoretical optimization landscapes. From engineering and finance to healthcare and beyond, ABC has showcased its prowess in solving real-world problems. Its simplicity, ease of implementation, and low computational requirements make it an attractive choice for various applications. We can derive a cumulative optimized solution using the generated equations given above for distributed and parallel computing. The future directions of the ABC algorithm promise even more exciting developments. Hybridization with other algorithms, adaptive strategies, and handling constraints are just a few areas where researchers can channel their efforts. Moreover, the algorithm's scalability can be enhanced through parallel and distributed computing, opening up possibilities for tackling larger and more intricate optimization tasks. The integration of cognitive-inspired modifications and a focus on real-world applications are crucial steps in making ABC more intelligent and applicable to practical scenarios. Fine-tuning algorithm parameters and establishing benchmarking standards will further contribute to the algorithm's effectiveness and widespread acceptance. In essence, the journey of solving with bees through the ABC algorithm represents a convergence of nature-inspired ingenuity and computational prowess. As researchers, practitioners, and the wider community collaborate, ABC is poised to continue its transformative impact, offering innovative solutions to the increasingly complex challenges faced across diverse fields. Whether in the intricate dance of optimization landscapes or the intricacies of real-world applications, the ABC algorithm remains a beacon of inspiration and problem-solving.

References

1. Bansal JC, Gopal A, Nagar AK (2018) Analysing convergence, consistency, and trajectory of artificial bee colony algorithm. IEEE Access 6:73593–73602
2. Chen M, Zhang H, Li X (2018) Application of artificial bee colony algorithm for timing of road entrance signal. In: Wang Y (ed) MATEC web of conferences, vol 232, p 03004
3. Somar S, Urazel B, Sahin YB (2023) A modified metaheuristic algorithm for a home health care routing problem with health team skill levels. Appl Soft Comput 148:110912–2
4. Bilim M, Karaboga D (2023) Improved Chernoff bound of Gaussian Q-function with ABC algorithm and its QAM applications to DB SC and MRC systems over Beaulieu-Xie channels. Phys Commun 1(58):102034–102044
5. Cui Y, Hu W, Rahmani A (2022) Fractional-order artificial bee colony algorithm with application in robot path planning. Eur J Oper Res
6. Cui Y (2023) Optimizing decision trees for English teaching quality evaluation (ETQE) using artificial bee colony (ABC) optimization. Heliyon 9(8):e19274–e19284
7. Li THS, Chiu HJ, Kuo PH (2022) Hepatitis C virus detection model by using Random Forest, Logistic-regression and ABC algorithm. IEEE Access 10:91045–91058
8. Taspinar N, Yildirim M (2015) A novel parallel artificial bee colony algorithm and its PAPR reduction performance using SLM scheme in OFDM and MIMO-OFDM systems. IEEE Commun Lett 19(10):1830–1833
9. Yang J, Li W, Shi X, Xin L, Yu J (2013) A hybrid ABC-DE algorithm and its application for time-modulated arrays pattern synthesis 61(11):5485–5495
10. Yang L, Sun X, Peng L, Yao X, Chi T (2015) An agent-based artificial bee colony (ABC) algorithm for hyperspectral image endmember extraction in parallel. IEEE J Sel Top Appl Earth Obs Remote Sens 8(10):4657–4664
11. Yao G, Wu Y, Huang X, Ma Q, Du J (2022) Clustering of typical wind power scenarios based on K-means clustering algorithm and improved artificial bee colony algorithm. IEEE Access 10:98752–98760
12. Han F, Li X, Qi S, Wang W, Shi W (2023) Reliability analysis of wind turbine subassemblies based on the 3-P Weibull model via an ergodic artificial bee colony algorithm. Probab Eng Mech 1(73):103476–103486
13. Akay R, Akay B (2020) Artificial bee colony algorithm and an application to software defect prediction. Model Optimiz Sci Technol 1:73–92
14. Tong J, Cao Z, Fu W (2023) Motion vector steganography algorithm of sports training video integrating with artificial bee colony algorithm and human-centered AI for web applications. J Intell Syst 32(1)

15. Xiang Y, Zhou Y, Tang L, Chen Z (2019) A decomposition-based many-objective artificial bee colony algorithm. IEEE Trans Cybern 49(1):287–300
16. Zhang Q, Bu X, Gao H, Li T, Zhang H (2023) A hierarchical learning based artificial bee colony algorithm for numerical global optimization and its applications. Appl Intell
17. Li G, Liu C, Wu L, Xiao W (2023) A mixing algorithm of ACO and ABC for solving path planning of mobile robot. Appl Soft Comput 1(148):110868–110878
18. Ewees AA, Abd ME, Mohammed, Khalil HA, Kim S (2020) Improved artificial bee colony using sine-cosine algorithm for multi-level thresholding image segmentation. IEEE Access 8:26304–26315
19. Jie Y (2020) Application of artificial bee colony algorithms in smart grid. J Phys: Conf Ser 1453(1):012083–012093
20. Habib AW, Junejo AK, Ismail MM, Hossen M, Jahangiri M et al (2022) Optimal planning of residential microgrids based on multiple demand response programs using ABC algorithm. IEEE Access 10:116564–116626
21. Ren X, Zhang H (2022) An improved artificial bee colony algorithm for model-free active noise control: algorithm and implementation. IEEE Trans Instrum Meas 1(71):1–11
22. Toktas A, Ustun D, Erdogan N (2020) Pioneer pareto artificial bee colony algorithm for three-dimensional objective space optimization of composite-based layered radar absorber. Appl Soft Comput 96:106696
23. Wang T, Fu L, Wei Z, Zhou Y, Gao S (2023) Unmanned ground weapon target assignment based on deep Q-learning network with an improved multi-objective artificial bee colony algorithm. Eng Appl Artif Intell 117:105612
24. Yildirim M (2020) ABC algorithm-based subblocks set design for OFDM with all index modulation. IEEE Commun Lett 24(12):2926–2930
25. Zhao F, Wang Z, Wang L, Xu T, Zhu N, Jonrinaldi (2023) An exploratory landscape analysis driven artificial bee colony algorithm with maximum entropic epistasis. Appl Soft Comput 137:110139
26. Lei D, Yang H (2022) Scheduling unrelated parallel machines with preventive maintenance and setup time: multi-sub-colony artificial bee colony. Appl Soft Comput 1(125):109154–109164
27. Sahin O, Akay B (2021) A discrete dynamic artificial bee colony with hyper-scout for RESTful web service API test suite generation. Appl Soft Comput 104:107246
28. Özkan R, Genç MS (2023) Aerodynamic design and optimization of a small-scale wind turbine blade using a novel artificial bee colony algorithm based on blade element momentum (ABC-BEM) theory. Energy Convers Manag 1(283):116937
29. Karaman A, Pacal I, Basturk A, Akay B, Nalbantoglu U, Coskun S et al (2023) Robust real-time polyp detection system design based on YOLO algorithms by optimizing activation functions and hyper-parameters with artificial bee colony (ABC). Expert Syst Appl 221:119741
30. Luo Q, Rao Y, Du B (2024) An effective discrete artificial bee colony for the rectangular cutting problem with guillotine in transformer manufacturing. Appl Soft Comput 1:111617–111627
31. Ganguli C, Shandilya SK, Izonin I (2023) Design and implementation of adaptive network stabilization based on artificial bees colony optimization for nature inspired cyber security. J King Saud Univ Sci 35(5):102713
32. Sutar M, Jadhav HT (2023) A modified artificial bee colony algorithm based on a non-dominated sorting genetic approach for combined economic-emission load dispatch problem. Appl Soft Comput 144:110433
33. Delgado-Osuna JA, Lozano M, García-Martínez C (2016) An alternative artificial bee colony algorithm with destructive–constructive neighbourhood operator for the problem of composing medical crews. Inf Sci 1(326):215–226
34. Zhang S, Xu S, Huang X, Zhang W, Chen M (2019) Networked correlation-aware manufacturing service supply chain optimization using an extended artificial bee colony algorithm. Appl Soft Comput 1(76):121–139
35. Djenouri Y, Belhadi A, Belkebir R (2018) Bees swarm optimization guided by data mining techniques for document information retrieval. Expert Syst Appl 94:126–136

36. Cui Y, Hu W, Rahmani A (2024) Multi-robot path planning using learning-based artificial bee colony algorithm. Eng Appl Artif Intell 1(129):107579–107589
37. Huda NHM (2022) Balance of exploration and exploitation in artificial bee colony for multi products inventory routing problem. Appl Math Sci 16(9):425–434
38. Bai Y, Zhang C, Bai W (2023) A two-level parallel decomposition-based artificial bee colony method for dynamic multi-objective optimization problems. Appl Soft Comput 1(147):110741–110751
39. Chen J, Cai Z, Heidari AA, Liu L, Chen H, Pan J (2023) Dynamic mechanism-assisted artificial bee colony optimization for image segmentation of COVID-19 chest X-ray 79:102485. https://www.sciencedirect.com/science/article/pii/S014193822300118X
40. Yang MC, Pan QK, Sang HY, Li WM, Wang YL (2023) A slack speed-up based discrete artificial bee colony algorithm for resource-constrained operating room scheduling problem. Comput Ind Eng 1(186):109760
41. Wang Z, Tang A (2023) Distribution network reconfiguration based on improved artificial bee colony algorithm
42. Zhang B, Che A, Wang Y (2024) Grid-based artificial bee colony algorithm for multi-objective job shop scheduling with manual loading and unloading tasks. Expert Syst Appl 1(245):123011–123021
43. Wang J, Liu Y, Rao S, Zhou X, Hu J (2023) A novel self-adaptive multi-strategy artificial bee colony algorithm for coverage optimization in wireless sensor networks. Ad Hoc Netw 1(150):103284–103294
44. Niu W, She W, Zhong L, Wang Y, Tian Z, Liu W (2023) Elite-centered artificial bee colony algorithm with extended solution boundary. Appl Soft Comput 1(148):110906–110916
45. Gul E, Toprak AN (2023) Contourlet and discrete cosine transform based quality guaranteed robust image watermarking method using artificial bee colony algorithm. Expert Syst Appl 212:118730
46. Li M, Chang CT, Liu Z (2024) A discrete artificial bee colony algorithm and its application in flexible flow shop scheduling with assembly and machine deterioration effect. Appl Soft Comput 1(159):111593–111603
47. Chen L, Wu T, Wang Z, Lin X, Cai Y (2023) A novel hybrid BPNN model based on adaptive evolutionary artificial bee colony algorithm for water quality index prediction. Ecol Ind 1(146):109882–109892
48. Comert SE, Yazgan HR (2023) A new approach based on hybrid ant colony optimization-artificial bee colony algorithm for multi-objective electric vehicle routing problems. Eng Appl Artif Intell 123:106375
49. Kolukisa B, Dedeturk BK, Hacilar H, Gungor VC (2024) An efficient network intrusion detection approach based on logistic regression model and parallel artificial bee colony algorithm. Comput Stand Interfaces 1(89):103808
50. Li M, Hei Y, Qiu Z (2017) Optimization of multiband cooperative spectrum sensing with modified artificial bee colony algorithm. Appl Soft Comput 57:751–759
51. Nikolić M, Šelmić M, Macura D, Ćalić J (2020) Bee colony optimization metaheuristic for fuzzy membership functions tuning. Expert Syst Appl 158:113601
52. Abu-Mouti FS, El-Hawary ME (2012) Overview of artificial bee colony (ABC) algorithm and its applications. In: 2012 IEEE international systems conference SysCon 2012, March 2012
53. Karas A, Ozcelik F (2021) Assembly line worker assignment and rebalancing problem: a mathematical model and an artificial bee colony algorithm. Comput Ind Eng 156:107195
54. Incorvaia DC, Dalrymple T, Huang ZY, Dyer FC (2022) Short- and long-term modulation of forager motivation by colony state in bumble bees. Anim Behav 1(190):61–70
55. Nikolić M, Rakas J, Teodorović D (2024) Formulation of the airport collaborative gate allocation problem and the bee colony optimization solution approach. Eng Appl Artif Intell 1(128):107433

Swarm Intelligence for Optimization: A Bee's-Eye View on Multi-objective and Dynamic Challenges

R. S. M. Lakshmi Patibandla, D. Madhusudhana Rao, and Y. Gokul

1 Introduction

Every industry you can think of has optimization issues, including finance, engineering design, resource allocation, and even personal decision-making. However, the majority of conventional optimization methods concentrate on just one goal, such as maximizing profit or minimizing expense. However, in the actual world, issues frequently include balancing numerous, conflicting agenda. This is where dynamic and multi-objective optimization becomes challenging. Consider creating an automobile. It should be inexpensive (low production cost), safe (strong, crash-resistant), fuel-efficient (low emissions), and comfortable (a roomy, smooth ride). These goals conflict often; giving one priority over another could jeopardize the other. In order to address this complexity, multi-objective optimization (MOO) finds Pareto optimal solutions, which are the best feasible compromises in situations where achieving one goal would not be at the expense of another. MOO algorithms use a variety of strategies to overcome these challenges [1]. Natural selection serves as the model for evolutionary algorithms, which generate a population of possible solutions and "breed" them according to how well they accomplish each goal. Multi-objective evolutionary algorithms (MOEAs) are excellent at identifying a variety of parametric optimal solutions, providing decision-makers with a selection of choices. Other techniques, such as goal programming and weighted sum, combine several objectives into a single function, but they need precise weighting modifications to produce the appropriate trade-offs. Another level of complexity is that real-world

R. S. M. Lakshmi Patibandla (✉)
Department of CSE, Koneru Lakshmaiah Education Foundation, Guntur, Andra Pradesh, India
e-mail: patibandla.lakshmi@gmail.com

D. Madhusudhana Rao
Department of CSE, Amrita Vishwa Vidyapeetham, Guntur, Andra Pradesh, India

Y. Gokul
Department of CSE, Vellore Institute of Technology, Vijayawada, Andra Pradesh, India

K. Raza (ed.), *Solving with Bees*, Springer Tracts in Nature-Inspired Computing,
https://doi.org/10.1007/978-981-97-7344-2_3

issues frequently evolve. Consider overseeing a traffic network where the best routes are influenced by weather, road conditions, and traffic patterns.

The process of making a system, design, or choice as functional or effective as feasible within predetermined parameters is known as optimization. It entails choosing the optimal choice from a set of workable solutions in mathematics and engineering by maximizing or decreasing factors, such as cost, efficiency, or performance. Business optimization can improve productivity, reduce expenses, increase profits, or improve processes. Supply chain optimization, for instance, seeks to reduce logistics expenses while guaranteeing prompt delivery. Using methods like machine learning, greedy algorithms, and dynamic programming, optimization in computer science improves software and algorithms' speed, performance, and use of resources. Optimization provides the best results across a wide range of domains by methodically weighing options and selecting the one that best satisfies goals. We frequently use the type of objective function and constraints that are linear or nonlinear, convex or non-convex to categorize problems. A variety of strategies, from straightforward ones like gradient descent to trickier ones like simulated annealing or genetic algorithms, can resolve these issues. Optimization finds practical applications in the following domains: manufacturing resource allocation, finance portfolio returns, delivery truck route optimization, microchip component layout design, and machine learning model parameter refinement for increased predictive accuracy. Beyond numerical issues, decision-making procedures heavily rely on optimization as a means of resolving issues involving a variety of trade-offs. Despite having a wide range of applications, optimization issues can vary widely in complexity, with some needing advanced computational and mathematical techniques to solve well. The ongoing advancements in optimization theory and methods highlight the basic importance of optimization in both theoretical and practical parts of science and industry, driving innovation and progress in many domains.

Dynamic optimization (DO) is required in this situation, where the algorithm continuously modifies the solution in response to these changes. DO algorithms to deal with this dynamism use different tactics. Future changes are predicted using predictive models, which enable the algorithm to make proactive adjustments to solutions [2]. Reactive methods adjust instantly in response to input from the surroundings. In DO, reinforcement learning—in which the algorithm gains knowledge from both successes and failures—is also showing promise. Even though they are strong, MOO and DO have significant obstacles. It can be computationally costly to find truly parametric optimal solutions, particularly when there are numerous objectives. Objective functions could be intricate and challenging to measure. In DO, poor forecasts or unforeseen adjustments may result in less-than-ideal choices. Notwithstanding these obstacles, developments in machine learning, processing speed, and algorithm design are opening the door to more potent and effective solutions. To ensure that algorithms are implemented ethically and responsibly and are tailored to specific issues, collaboration between mathematicians, engineers, and domain experts is essential. MOO and DO are essential tools for negotiating the complexity of real-world problems, from balancing automobile design to optimizing traffic flow. Through the resolution of these complex and ever-changing issues, we can improve decision-making, maximize results, and build a more effective and sustainable future. Nature is a team player [3]. Collaborative behavior frequently results in amazing feats, from ant colonies painstakingly creating intricate networks to flocks of birds determining magnificent formations. Swarm Intelligence (SI), a fascinating branch of artificial intelligence, is inspired by this complex dance of coordinated movement and decision-making.

SI algorithms provide unique solutions with a novel "bee's-eye view" by emulating the collaboration and communication seen in natural swarms [4]. This allows them to handle complicated problems without centralized control. SI algorithms use the combined knowledge and interactions of many simple agents rather than depending on a single, powerful intellect. Similar to how bees use their nuanced dances to alert their hive to potential food sources, SI agents also use local interactions and straightforward rules to share information. By fostering adaptability and robustness, this decentralized strategy enables the swarm to overcome challenges that a single, centralized intellect might find difficult to overcome and adapt to changing settings [5].

There are several variations of SI algorithms, each one inspired by a distinct natural phenomenon. Ant colony optimization (ACO) guides virtual "ants" through a problem space to find optimal solutions, replicating how ants locate the shortest path to food. Particle swarm optimization (PSO) modifies the paths of its agents according to their individual and collective knowledge, simulating the coordinated flight of bird flocks. Other algorithms take their cues from the collective feeding behaviors of bats, fish schools, and even the movement of bacteria. SI's inherent decentralization provides a distinctive angle on problem-solving [6]. Consider negotiating a maze of obstacles. Some may find it difficult to see the whole structure and determine the best course of action from a centralized "top-down" viewpoint. The "bee's-eye view" of SI, on the other hand, enables agents to investigate multiple paths at once, sharing knowledge and progressively settling on the optimal one. This all-encompassing method can be very helpful when solving issues in dynamic situations and complex interactions [7].

SI is becoming well-known in many different fields. It helps build effective telecommunication networks, optimizes routing in transportation networks, and even supports robotics by allowing swarms of robots to work together on tasks. Its potential in financial forecasting, scheduling, and drug development is being investigated by researchers. SI algorithms' potential uses will undoubtedly grow as computing power and algorithmic sophistication both advance. Even with its fascinating potential, SI has drawbacks. It can be difficult to create efficient communication channels for complicated issues. It takes careful adaptation to strike a balance between the two forces of exploration and exploitation—finding new options and improving established solutions. Responsible attention must be given to ethical issues and possible biases resulting from the underlying data used to train SI algorithms [8]. Still, researchers and developers are continually inspired by the tremendous potential of SI. SI provides a useful viewpoint on problem-solving, inventive solutions, and a more cooperative future for artificial intelligence by utilizing collective wisdom. The "bee's-eye view" may revolutionize how we tackle some of our toughest problems.

2 Bee-Inspired Algorithms for Optimization

2.1 Key Principles of Swarm Intelligence: Collective Decision-Making, Communication, and Adaptation

The complex network of life on Earth exhibits astounding achievements of collective intelligence. These considerations, which range from busy ant colonies creating intricate networks to captivating flocks of birds engaging in coordinated movements, are frequently the result of the ostensibly straightforward interactions of many individuals. This fascinating dance between cooperation and judgment has given rise to swarm intelligence (SI), a potent subset of artificial intelligence. Now let us explore the three main ideas that form the foundation of this exciting field.

2.1.1 Collective Decision-Making: Wisdom of the Crowd in Action

Unlike conventional AI techniques, which depend on a single, potent mind, swarm intelligence uses several intelligences. Rather, it makes use of the interactions and cumulative knowledge of many simple agents. Imagine a busy beehive where individual bees use nuanced dances to guide the swarm's exploring efforts and transmit the location of food sources [9]. Similarly, without centralized control, SI algorithms use basic rules and local interactions between agents to allow them to collectively find optimal solutions. By enabling resilience and flexibility, this decentralized method enables the swarm to overcome constraints that could impede a single intellect and adapt to dynamic circumstances.

2.1.2 Communication: The Language of the Swarm

Effective communication is essential to any successful team effort. This corresponds to several agent-to-agent information exchange techniques in SI. ACO involves virtual "ants" that leave pheromone trails along possible routes, directing other ants toward possible food sources. PSO agents exchange their optimal placements with one another, guiding the swarm as a whole towards more advantageous outcomes [10]. Other algorithms are influenced by bacterial chemotaxis, fish schooling communication mechanisms, and even bat echolocation tactics. Agents can pool their expertise, learn from each other's successes and errors, and jointly improve their methods thanks to these various communication protocols.

2.1.3 Adaptation: Navigating the Dynamic Maze

Problems in the real world are rarely static. Adaptation is essential for SI algorithms to succeed in uncertain situations. SI agents constantly adapt their behavior based on their experiences and interactions, much like how ants dynamically change their foraging routes in response to changing food availability. Mechanisms like memory, which allow agents to learn from their prior successes and errors, are the source of this flexibility. PSO particles continuously improve their solutions by adjusting their velocities according to their individual and collective optimal locations [11]. Pheromone trails in ACO dynamically vanish over time, promoting the discovery of fresh, promising avenues while maintaining the swarm's attention on previously found, effective solutions. This ongoing balance between exploration and exploitation makes sure the swarm can both identify the best solutions and adjust to environmental changes.

These basic concepts enable SI to address a wide range of problems in numerous fields. It plans effective communication networks, streamlines urban traffic flow, and even directs teams of robots to work together on projects. Researchers are investigating its potential in medication discovery, scheduling issues, and even financial predictions. SI applications are certain to expand as algorithms advance and processing capability increases.

2.2 *Challenges*

Even with its tremendous promise, SI has difficulties. It can be difficult to create efficient communication channels for complicated issues. It takes careful adaptation to strike a balance between the two forces of exploration and exploitation—finding new options and improving established solutions. Due diligence must be given to ethical issues arising from possible biases in the data utilized to train SI algorithms [12].

Even so, researchers and developers are continually inspired by the tremendous potential of SI. SI provides a distinct viewpoint on problem-solving, offering creative solutions and a more collaborative future for artificial intelligence by exploiting the combined knowledge of multiple agents. The "wisdom of the crowd" may hold previously undiscovered answers to some of our most urgent problems as we continue to unravel the complex dance of the swarm.

2.2.1 Exploring Prominent Bee-Inspired Algorithms: ABC, ACO, PSO

Technology and engineering have plenty to learn from nature, and bees' hardworking lives are no different. The concepts of bee behavior are included in several well-known swarm intelligence algorithms, highlighting the strength of cooperation and collective intelligence. Let us examine ABC, ACO, and PSO, three prominent bee-inspired algorithms [13].

ABC—Sharing the Nectar Harvest: Employed bees, observer bees, and scout bees are the three categories of agents used by ABC, which is modeled after the honeybees'

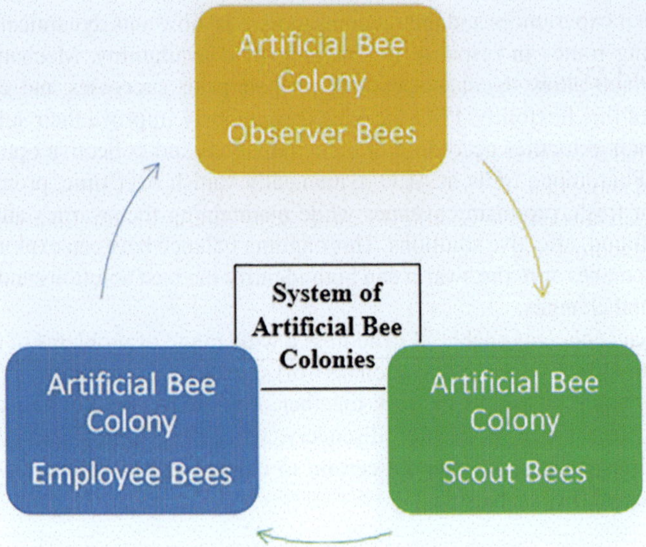

Fig. 1 System of different types of Artificial Bee Colonies

foraging techniques. Working on well-known food sources or solutions, employed bees use a "dance" language (a probabilistic selection method) to communicate their qualities to other bees. Based on these data and the success rates of the employed bees, onlooker bees select their food sources. Scout bees widen their search area by investigating novel food sources and possible remedies to prevent stagnation. Up until an optimal or suitable level is reached, the algorithm iteratively refines these solutions, giving preference to those with higher quality (fitness). Figure 1 represents the system of different types of bee colonies like Observer Bees, Employee Bees, and Scout Bees.

ACO—Following the Pheromone Trails: ACO is similar to how ants forage, leaving behind pheromones that direct other ants towards possible food sources. The system creates "pheromone trails" on viable routes while artificial ants move through a virtual environment that represents the issue region. These trails' intensity serves as a metric for the path's perceived quality, drawing additional ants and highlighting effective paths. Eventually, the colony shows remarkable problem-solving skills as it settles on the optimal or easiest route to reach its objective.

PSO—Learning from the Collective Flight: Inspired by the coordinated flight patterns of animal flocks, PSO employs particles that stand in for possible solutions inside a search space. Based on the best-known solution for the swarm as well as each particle's individual best solution, each particle has a position and velocity. Particles move and adjust their positions in response to these forces, searching the search space and settling in on areas of potential interest. Iteratively refining its grasp of the problem and converging towards an optimal solution, the swarm learns collectively and shares information [14].

2.3 *Similarities and Differences*

All three algorithms—ACO, ABC, and PSO—are inspired by nature, yet they exhibit notable differences. ACO primarily relies on pheromone trails for navigation, whereas ABC uses a blend of exploitation by employed bees and exploration by scout bees. PSO, on the other hand, emphasizes particle-to-particle communication and social learning. In terms of information sharing, PSO depends on both the best experiences within the swarm and individual experiences, ACO uses pheromone trails, and ABC employs a probabilistic dance language. The applications of these algorithms also vary: PSO is often used in data clustering and engineering design, ACO thrives in pathfinding and routing problems, and ABC excels in numerical function optimization and scheduling [15].

2.3.1 Beyond the Hive: The Buzzing Future of Swarm Intelligence

These bee-patterned algorithms represent only a fraction of the intriguing field of swarm intelligence. Ongoing research continues to push boundaries by developing new algorithms and enhancing existing ones [16]. As computing power increases and our understanding of natural swarming behavior deepens, swarm intelligence holds significant potential for innovative solutions to complex problems across various sectors. Perhaps a touch of honeyed inspiration will spark the next breakthrough in AI.

2.4 *Benefits of Using Bee-Inspired Algorithms for Multi-objective Problems*

Multi-objective problems (MOPs) pose unique challenges by requiring the balancing and optimization of several, often conflicting, objectives. Bee-inspired algorithms, leveraging the collective intelligence and effective foraging techniques of bees, offer several advantages in addressing these challenges.

2.4.1 Balancing Exploration and Exploitation

For MOPs, it is crucial to balance the refinement of existing solutions with the search for a variety of alternatives. Bee-inspired algorithms excel in this aspect. ABC utilizes scout bees for exploration and employed bees for exploitation. ACO exploits successful trails while pheromone evaporation promotes the discovery of new routes. PSO guides particles to promising areas based on their personal best positions and the swarm's best positions, balancing individual exploration with group wisdom. This innate capability allows them to effectively navigate the complex trade-offs inherent in MOPs.

2.4.2 Distributed and Robust Solutions

Unlike conventional, centralized methods, bee-inspired algorithms depend on the cooperation of multiple simple agents, offering several advantages. Robustness is ensured as the failure of one agent does not compromise the entire algorithm, since other agents can compensate. Adaptability allows agents to quickly adjust their behaviors based on feedback and local interactions, making the algorithms suitable for dynamic environments. Parallelization enables these algorithms to be efficiently parallelized across multiple processors, leveraging computing power to solve problems faster. These features make them particularly well-suited for MOPs requiring diverse, flexible solutions.

2.4.3 Efficient Information Sharing

Bees effectively share information about potential solutions through their "dance language" or pheromone trails, guiding others to successful strategies. This efficient communication translates to bee-inspired algorithms [17]. ABC uses a probabilistic dance language based on solution quality to guide other bees. ACO directs ants to superior paths through the strength of pheromone trails, promoting group learning. PSO enables particles to move toward optimal regions using collective knowledge about personal and swarm-best locations. This efficient information exchange enables rapid convergence of promising solutions within the multi-objective space.

2.4.4 Beyond the Benefits

Despite their numerous advantages, bee-inspired algorithms have limitations. Parameter tuning can be challenging, as finding the optimal values for algorithm parameters is not straightforward. Computational complexity may also increase with the number of agents in large-scale MOPs, raising computational costs [18–20]. Additionally, while these algorithms perform well in practice, a deeper theoretical understanding is still needed. Overall, bee-inspired algorithms offer a robust and adaptable toolkit for solving multi-objective problems. Their inherent strengths in balancing exploration and exploitation, providing distributed and robust solutions, and promoting efficient information sharing make them valuable tools for researchers and practitioners seeking innovative solutions to complex problems involving multiple, often conflicting, objectives.

3 Dynamic Optimization with Swarm Intelligence

3.1 Adapting Algorithms to Changing Environments

Imagine navigating an ever-evolving environment where paths appear and disappear, bottlenecks arise, and the ultimate goal is in flux. Conventional optimization algorithms, designed for static problems, struggle in such dynamic settings. Swarm intelligence and dynamic optimization offer effective solutions, allowing algorithms to adapt to constantly changing conditions.

3.2 The Challenge of Constant Change

Real-world problems are seldom static. Financial markets fluctuate, traffic patterns shift, and weather conditions vary. In these scenarios, optimizing systems requires ongoing adjustments, rendering static, one-size-fits-all approaches inadequate. Dynamic optimization (DO) addresses this challenge by updating solutions in real time based on changes in inputs and the environment.

3.3 Swarm Intelligence: Wisdom of the Crowd in Action

Swarm optimization (SO) leverages principles of swarm intelligence, inspired by the cooperative behavior of social insects like bees and ants. These algorithms employ numerous simple agents that communicate and interact to solve problems that centralized control cannot. Their decentralized nature makes them inherently robust and adaptive, crucial for navigating unpredictable environments.

3.4 Swarm-Based Strategies for Dynamic Adaptation

Swarm-based strategies for dynamic adaptation include predictive models, which forecast future changes and enable proactive solution adjustments [21]. Reactive strategies allow agents to dynamically alter their behavior and search processes based on real-time environmental feedback. Additionally, reinforcement learning can be utilized, enabling agents to learn from their successes and failures in real time by adapting their strategies in response to shifting rewards and penalties.

3.5 Benefits of Swarm-Based Dynamic Optimization

Swarm-based dynamic optimization offers several benefits. Adaptability allows SO algorithms to continuously refine solutions, making them resilient to sudden changes. Decentralization ensures that the failure of individual agents does not compromise the entire algorithm. These algorithms also balance the exploration of new possibilities with exploitation of existing solutions, which is essential in dynamic settings. Furthermore, SO algorithms are easily parallelizable, allowing them to leverage additional computing power for faster adaptation [22].

4 Real-World Applications

4.1 SO in Dynamic Domains

Swarm Optimization (SO) is effectively applied across various dynamic domains. In traffic signal optimization, SO adjusts signal timings based on real-time traffic data, improving flow and reducing congestion. In financial trading, it modifies investment strategies in response to market changes, enhancing profitability []. For robot path planning, SO helps robots adaptively avoid obstacles and follow shifting objectives. In smart grids, SO optimizes electricity distribution by balancing demand with the availability of renewable energy.

4.2 Challenges and the Future

While promising, SO faces several challenges. Large-scale implementations may require significant computational resources due to computational complexity. Optimizing algorithm parameters for specific dynamic environments can be challenging and time-consuming. Ethical considerations must also be addressed, particularly biases in the data used for training SO algorithms. Despite these challenges, research in SO continues to advance, pushing the boundaries of dynamic optimization. By harnessing the collective intelligence of swarms, SO offers a powerful tool for tackling complex, ever-changing challenges, paving the way for a more adaptable and responsive future.

4.3 Handling Real-Time Data and Feedback

4.3.1 Navigating the Rapids: Dynamic Optimization with Swarm Intelligence in Real-Time Environments

Imagine navigating a fast-flowing river where rocks emerge, rapids roar, and the current shifts unpredictably. To succeed, you must respond swiftly, adjusting your course based on real-time information. This dynamic, quick-thinking decision-making process mirrors the challenges in various real-world scenarios, such as energy grid optimization and traffic flow management. Dynamic optimization using Swarm Intelligence provides a powerful approach for managing real-time data and feedback for optimal decision-making [23, 24].

4.3.2 The Real-Time Challenge

In real-time environments where data streams continuously and conditions change constantly, traditional optimization techniques often falter. Consider traffic signal optimization: algorithms need to adjust in real time to changing traffic patterns, accidents, and weather conditions. Static data analysis won't suffice. SO tackles this challenge with key principles:

- **Decentralized Collaboration**: SO relies on multiple simple agents interacting without a single leader, akin to how social insects like ants solve problems without centralized control. This decentralization allows SO to quickly adapt to changes and remain resilient to failures.
- **Constant Feedback Loop**: SO algorithms are not one-shot solutions. They continuously receive and process real-time data, updating their strategies based on the latest information. Just as ants would communicate a better food source to the colony, SO algorithms dynamically adapt to new data.
- **Learning and Adaptation**: SO can incorporate reinforcement learning, allowing agents to adjust their behaviors based on real-time successes and failures. This continuous learning process helps agents optimize their performance in a constantly changing environment.

4.3.3 Unlocking Real-Time Optimization

Combining these principles offers significant benefits for real-time applications:

- **Quick Response**: Decentralized processing and real-time feedback create agile systems that swiftly adjust to changing conditions.
- **Robustness**: The failure of individual agents does not compromise the system, ensuring continued optimization.
- **Scalability**: SO algorithms can efficiently handle large real-time data streams due to their inherent parallelizability.

4.3.4 Real-World Applications

SO is gaining traction in numerous fields:

- **Traffic Signal Optimization**: Dynamically adjusting signal timings to improve traffic flow and reduce congestion based on real-time data.
- **Financial Trading**: Adjusting strategies in real-time according to market swings and sentiment analysis.
- **Robot Control**: Enabling robots to navigate dynamic environments by responding to sensor inputs and altering their paths.
- **Smart Grids**: Efficiently allocating resources and distributing electricity in response to current demand and the availability of renewable energy.

4.3.5 Navigating the Challenges

Despite its potential, SO faces challenges:

- **Computational Complexity**: Processing large datasets in real time requires substantial computational power.
- **Parameter Tuning**: Optimizing algorithm parameters for specific conditions necessitates careful calibration.
- **Ethical Considerations**: Addressing biases in the data used to train SO algorithms is crucial for ensuring fairness and ethical decision-making.

4.3.6 The Future of Real-Time Optimization

Research in SO continues to address these challenges, and its potential is expanding. By leveraging real-time data and the collective intelligence of swarms, SO provides a powerful tool for navigating the ever-changing rapids of our dynamic world. This technology holds tremendous promise for real-time decision-making, streamlining complex processes, and ultimately creating a more adaptable and responsive future.

5 Practical Applications and Case Studies

The algorithms replicate the hive's efficient foraging strategies and collective intelligence of the beehive and have evolved into something more than just a fascinating field of study. These days, they are actively addressing difficult problems in many different areas, proving their usefulness in many real-world contexts. Honeycomb optimization has played a significant role in the development of strong yet lightweight aircraft components in the field of engineering design, drawing inspiration from honeycomb structure. Because of this, Honeywell has been able to reduce fuel consumption by a substantial amount and save millions of dollars. ACO has optimized data center climate control architectures, resulting in significant energy savings and reduced carbon emissions.

The ABC algorithm optimizes courier delivery routes in logistics and scheduling, resulting in lower costs and faster delivery times. PSO in task planning has led to increased output and decreased idle time in sophisticated manufacturing settings. Communication networks use ACO for resource allocation to distribute bandwidth efficiently, maximizing resource consumption and improving network efficiency. In a similar vein, agricultural systems use the bee algorithm to plan irrigation, which increases crop yields while decreasing water waste. Bee-inspired algorithms have proven their adaptability in many other areas, not just these. Portfolio management and investment strategy optimization are two applications of these algorithms in financial trading. They are useful in the medical industry for evaluating medical pictures and making diagnoses. In the field of robotics, they help machines navigate complicated surroundings and carry out tasks more effectively. The inherent features of algorithms inspired by bees are the key to their success. Because there is no central authority or point of failure in a decentralized system, it is resilient and flexible. Collective intelligence, the result of agents working together and sharing data, allows for more efficient search and problem-solving. On top of that, these algorithms are good at both discovering new answers and optimizing the ones they already have. Table 1 represents unconstrained benchmark functions, used parameters for simulation shown in Table 2. Results are displayed in Table 3.

Despite their successes, there are obstacles to overcome for algorithms inspired by bees. Tuning parameters can be challenging, and large-scale applications might necessitate a lot of computing power. Understanding these algorithms and how they solve problems can be challenging due to their interpretability. However, researchers continue to study these problems, indicating a promising future for algorithms inspired by bees. More and more remarkable real-world applications are on the horizon, demonstrating the crucial role that algorithms and computer power play in solving complicated issues in a wide range of industries. The growing "buzz" about these algorithms is a reflection of their increasing influence, thanks to the collective intelligence of the beehive.

Table 1 Unconstrained benchmark functions

Function name	Mathematical notation	Search range	Maximum velocity
Sphere function	$f_1(x) = \sum_{i=1}^{n} x_i^2$	$(-5.13,5.13)^n$	5.12
RosenBrock function	$f_2(x) = \sum_{i=1}^{n-1}[100(x_i^2 - x_{i+1})^2 + (x_i - 1)^2]_1$	$(-5,10)^n$	10
Rastrigrin function	$f_3(x) = 10n + \sum_{i=1}^{n}[x_i^2 - 10cos(2\Pi x_i)]_1$	$(-5.13,5.13)^n$	5.12
Griewank function	$f_4(x) = \sum_{i=1}^{n} \frac{x_i^2}{400} - \prod_{i=1}^{n} cos(\frac{x_i}{\sqrt{i}})+1$	$(-601,601)^n$	600

Table 2 Used parameters for simulation

Parameter set	Population	Dimension	Generation
A1	20	20	2000
		40	4000
		60	6000
A2	40	20	2000
		40	4000
		60	6000
A3	60	20	2000
		40	4000
		60	6000
A4	80	20	2000
		40	4000
		60	6000

Table 3 Results for benchmark functions

Benchmark functions	Methodology	Mean	Standard deviation	Count of function calls
A1	ACO	0.14	16.08	19,989
	ABC	0.34	0.67	19,767
	PSO	0.17	46.34	20,000
A2	ACO	0.34	15.18	19,899
	ABC	0.24	0.54	29,665
	PSO	0.27	43.28	40,000
A3	ACO	0.15	16.78	29,989
	ABC	0.38	0.62	29,787
	PSO	0.18	45.76	60,000
A4	ACO	0.13	15.98	39,989
	ABC	0.35	0.72	39,767
	PSO	0.16	47.09	60,000

6 Conclusion and Future Scope

This chapter explores how nature's wisdom can solve tough optimization problems. It delves into swarm intelligence algorithms, inspired by bees' cooperative behavior, and shows how these algorithms can tackle dynamic challenges. Real-world applications demonstrate their versatility and efficiency in handling complex, conflicting objectives. Swarm intelligence, driven by collective hive intelligence, has the potential to revolutionize optimization. Enhanced by computing power, customization, and machine learning, it promises quicker, more precise solutions in fields like healthcare and finance. This era brings advancements in sustainability and artificial general intelligence research, but ethical and diverse growth is essential as swarm intelligence reshapes our world.

References

1. Qawqzeh Y, Alharbi MT, Jaradat A, Abdul Sattar KN (2021) A review of swarm intelligence algorithms deployment for scheduling and optimization in cloud computing environments. PeerJ Comput Sci 7. https://doi.org/10.7717/peerj-cs.696
2. Alsaidy SA, Abbood AD, Sahib MA (2020) Heuristic initialization of PSO task scheduling algorithm in cloud computing. J King Saud Univ. https://doi.org/10.1016/j.jksuci.2020.11.002. Epub ahead of print Nov 13 2020.
3. Aslan S (2019) A transition control mechanism for artificial bee colony (ABC) algorithm. Comput Intell Neurosci. 5012313. https://doi.org/10.1155/2019/5012313
4. Aslan S, Karaboga D (2020) A genetic Artificial Bee Colony algorithm for signal reconstruction based big data optimization. Appl Soft Comput J 88:106053–106053. https://doi.org/10.1016/j.asoc.2019.106053
5. Beed R, Roy A, Bhattacharya D (2019) A hybrid multi-objective tour route optimization algorithm based on particle swarm optimization and artificial bee colony optimization. Comput Intell 88:884–909
6. Dai W, Liang L, Zhang B (2020) Firefly optimization algorithm for the prediction of uplift due to high-pressure jet grouting. Adv Civ Eng 8833784. https://doi.org/10.1155/2020/8833784
7. Dai X, Long S, Zhang Z, Gong D (2019) Mobile robot path planning based on ant colony algorithm with a * heuristic method. Front Neurorobot 13:15. https://doi.org/10.3389/fnbot.2019.00015
8. Dokeroglu T, Sevinc E, Cosar A (2019) Artificial bee colony optimization for the quadratic assignment problem. Appl Soft Comput J 76:595–606. https://doi.org/10.1016/j.asoc.2019.01.001
9. Ewees AA, Al-qaness M, Elaziz MA (2021) Enhanced salp swarm algorithm based on firefly algorithm for unrelated parallel machine scheduling with setup times. Appl Math Modell 94:285–305. https://doi.org/10.1016/j.apm.2021.01.017
10. Farahlina Johari N, Mohd Zain A, Haszlinna Mustaffa N, Udin A (2017) Machining parameters optimization using hybrid firefly algorithm and particle swarm optimization. J Phys. Conf Series 892(1):012005. https://doi.org/10.1088/1742-6596/892/1/012005
11. Farid M, Latip R, Hussin M, Hamid N (2020) A survey on QoS requirements based on particle swarm optimization scheduling techniques for workflow scheduling in cloud computing. Symmetry (Basel) 12(4). https://doi.org/10.3390/SYM12040551
12. Hussein BM, Jaber AS (2020) Unit commitment based on modified firefly algorithm. Meas Control (United Kingdom) 53(3–4):320–327. https://doi.org/10.1177/0020294019890630
13. Khan A, Hizam H, Wahab N, Othman ML (2020) Optimal power flow using hybrid firefly and particle swarm optimization algorithm. PLOS ONE 15(8):1–21. https://doi.org/10.1371/journal.pone.0235668
14. Kruekaew B, Kimpan W (2020) Enhancing of artificial bee colony algorithm for virtual machine scheduling and load balancing problem in cloud computing. Int J Comput Intell Syst 13(1):496–510. https://doi.org/10.2991/ijcis.d.200410.002
15. Li X, Gao R, Yu L, Li N, Liu J, Wang Y, Feng E (2021) Memetic strategy of particle swarm optimization for one-dimensional magnetotelluric inversions. Mathematics 9(519):1–22
16. Nayak J, Naik B, Dinesh P, Vakula K, Dash PB (2020) Firefly algorithm in biomedical and health care: advances, issues and challenges. SN Comput Sci 1(6):311. https://doi.org/10.1007/s42979-020-00320-x
17. Pradhan A, Bisoy SK (2020) A novel load balancing technique for cloud computing platform based on PSO. J King Saud Univ. https://doi.org/10.1016/j.jksuci.2020.10.016. Epub ahead of print Oct 22 2020

18. Shahabi Sani N, Manthouri M, Farivar F (2020) A multi-objective ant colony optimization algorithm for community detection in complex networks. J Ambient Intell Human Comput 11(1):5–21. https://doi.org/10.1007/s12652-018-1159-7
19. Sheoran S, Mittal N, Gelbukh A (2020) Artificial bee colony algorithm in data flow testing for optimal test suite generation. Int J Syst Assur Eng Manag 11(2):340–349. https://doi.org/10.1007/s13198-019-00862-1
20. Tabrizchi H, Kuchaki Rafsanjani M, Balas VE (2021) Multi-task scheduling algorithm based on self-adaptive hybrid ICA–PSO algorithm in cloud environment. Adv Intell Syst Comput 1222:422–431. https://doi.org/10.1007/978-3-030-52190-5_30
21. Tamura Y, Sakiyama T, Arizono I (2021) Ant colony optimization using common social information and self-memory. Complexity. https://doi.org/10.1155/2021/6610670
22. Windarto W, Eridani E (2020) Comparison of particle swarm optimization and firefly algorithm in parameter estimation of lotka-volterra. AIP Conference Proceedings 2268(1):050008. https://doi.org/10.1063/5.0017245
23. Zhang G, Wang H, Zhao W, Guan Z, Li P (2021) Application of improved multi-objective ant colony optimization algorithm in ship weather routing. J Ocean Univ China 20:45–55. https://doi.org/10.1007/s11802-021-4436-6
24. Zhao R, Liu Q, Li C, Wang Y, Dong D (2020) Performance comparison and application of SI algorithms in crowd evacuation. In: ACM international conference proceeding series, pp 47–51

Applications of Artificial Bee Colony

Artificial Bee Colony Algorithms in Control Systems, Robotics, and Automation

Aditya Narayan Singh, Yagyesh Godiyal, Vibha Saroha, Yash Vardhan Singh, and Nripendra Kumar Singh

1 Introduction

The Artificial Bee Colony (ABC) is an optimization algorithm inspired by the foraging behavior of honeybees. It was proposed by Karaboga in 2005 and has since been applied to various real-world problem-solving scenarios due to its flexibility, simplicity, and efficiency in finding high-quality solutions, including system control, robotics, and automation. The ABC algorithm simulates the intelligent foraging behavior of honeybee swarms, comprised of three types of bees: *Employed Bees*, *Onlooker Bees*, and *Scout Bees* [14].

Employed Bees Phase: First, the algorithm generates a random population of solutions. Employed bees visit food sources (candidate solutions) and use an objective function to assess the quality of each as they search the search space. Each employed bee uses a local search strategy (such as a random perturbation) linked to a specific solution to investigate an adjacent solution within the solution space. Employed bees assess the quality of surrounding solutions and, depending on how fit the new solution is, decide whether to memorize it or forsake their present solution [15].

Onlooker Bees Phase: Observer bees watch the employed bees' dances, which indicate the caliber of the food supplies. An observer bee's likelihood of choosing a specific food source is directly correlated with its quality. Based on the information that employed bees provide, onlooker bees select a food source and investigate the appropriate solution, assessing its fitness and maybe refining it through a local search.

Scout Bees Phase: It is the duty of scout bees to investigate potential new food sources in the search area. By exploring uncharted or less promising areas at random, they add diversity to the population. A scout bee is sent out to randomly create a new

A. N. Singh · Y. Godiyal · V. Saroha · Y. V. Singh · N. K. Singh (✉)
Department of Computer Applications, Galgotias College of Engineering and Technology, Knowledge Park I, Greater Noida, Uttar Pradesh 201310, India
e-mail: nripendrakr123@gmail.com

K. Raza (ed.), *Solving with Bees*, Springer Tracts in Nature-Inspired Computing,
https://doi.org/10.1007/978-981-97-7344-2_4

solution if a food source (solution) is not improved after a predetermined number of repetitions (cycles) without improvement.

Termination Criteria: The ABC algorithm repeatedly goes through the stages mentioned above. This criterion could be exceeding a predetermined improvement threshold, the maximum number of iterations, or achieving a desirable level of solution quality.

Parameter Tuning: The number of bees, the trial limit, the number of cycles without improvement before abandoning, and the selection methods are some of the factors that affect how well the ABC algorithm performs.

The ABC algorithm has been extensively utilized in control systems, robotics, and automation across different disciplines since its establishment. The main advantage of ABC algorithms is their simplicity and effectiveness in tackling intricate optimization issues, which has resulted in their widespread application in the development of intelligent control methods, optimization of robotic pathways, and improvement of automation processes.

The utilization of ABC algorithms in control systems has shown substantial enhancements in the accuracy and stability of controllers. For example, the utilization of ABC optimization in the design of PID controllers has led to improved performance in a range of industrial applications, including steel strip deviation control systems [36]. The algorithm's capacity to dynamically adjust controller settings enables the implementation of more adaptable and resilient control mechanisms, which are essential for ensuring system stability in the face of changing operational conditions [8]. In addition, the ABC algorithm has been utilized to optimize nonlinear predictive control systems, demonstrating its efficacy in intricate control settings [2].

The application of ABC algorithms has significantly enhanced the field of robotics and automation. ABC algorithms have been employed in the field of robotic path planning to create effective and safe routes for mobile robots, resulting in enhanced operational efficiency and safety [21, 24]. In automated manufacturing systems, ABC algorithms have been utilized to improve job scheduling and coordinate robotic arms, resulting in enhanced production processes that are more efficient and streamlined [31]. By incorporating fuzzy logic, the integration of ABC has significantly improved the functionality of robotic control systems. This advancement allows for real-time machine learning control of articulated robotic manipulators [11].

ABC algorithms have a wide range of applications beyond the usual control and robotic fields. An example of this is the algorithm's adaptation to enhance the efficiency of robotic manipulators, as evidenced by the creation of interval type-2 fuzzy PID [20]. The ABC algorithm was employed in a different context to improve the speed control of robotic displacement, resulting in enhanced precision and efficiency of robotics [35]. The developments highlight the wide range of applications and efficiency of ABC algorithms in enhancing several elements of control systems, robotics, and automation.

2 Applications and Extensions

ABC algorithms have been successfully applied to various optimization problems, including function optimization, parameter tuning, feature selection, clustering, classification, and scheduling, across various domains such as automation, robotics, control systems, finance, healthcare, and logistics. The following are some scenarios and fields in which ABC algorithms are applied:

Parameter Optimization in Control Systems: To improve system performance, stability, and robustness, ABC algorithms can optimize the parameters of control systems, such as PID, fuzzy, or neural network controllers.

Trajectory Planning and Path Optimization for Robotics: ABC algorithms can be used to plan robot trajectories and optimize robot paths, taking into account variables like job completion time, energy efficiency, and obstacle avoidance.

Robot Swarm Coordination: By maximizing a swarm of robots' behaviors, paths, and communication techniques to effectively complete group tasks, ABC algorithms can promote cooperation and coordination among the robots.

Resource Allocation and Scheduling in Automation: To increase productivity, cut costs, and meet deadlines, ABC algorithms can optimize resource allocation and scheduling activities in automated systems, including supply chain management, traffic control, and industrial processes.

Optimization of Neural Networks and Machine Learning Models: In artificial intelligence, ABC algorithms are used to make neural networks and other machine learning models work better for tasks like classification, regression, and reinforcement learning.

Nature-Inspired Optimization: Honeybee colonies' efficient systems for resource exploitation and exploration serve as an inspiration for ABC algorithms. This methodology, which draws inspiration from nature, offers a distinctive viewpoint on optimization and frequently produces creative ideas that more conventional approaches could miss.

Population-Based Search: ABC methods employ a parallel exploration of the solution space by means of a population of candidate solutions, symbolized as bees. The method can preserve diversity and simultaneously explore different portions of the search space thanks to this population-based approach, which might be helpful for locating high-quality solutions to challenging, multi-dimensional optimization problems.

Global and Local Exploration: ABC algorithms strike a balance between local exploitation, which involves improving solutions close to potential regions, and global exploration, which involves looking for promising regions throughout the whole solution space. This equilibrium enables the algorithm to efficiently investigate various areas within the search space while progressively approaching ideal answers.

Adaptability and Robustness: ABC algorithms perform well in a variety of optimization scenarios despite being comparatively straightforward and simple to use. They can be used in situations where the problem domain is complicated or dynamic because they don't require gradient information or problem-specific parameter tweaking.

Efficiency and Scalability: ABC algorithms can handle optimization problems with big search areas or real-time restrictions because of their computational efficiency and scalability. They can manage issues with stochastic disturbances, high-dimensional parameter spaces, and nonlinear dynamics that are frequently seen in automation, robotics, and control systems.

Integration with Complex Systems: ABC algorithms offer optimization capabilities for tasks including parameter tuning, trajectory planning, resource allocation, and scheduling. They may be easily incorporated into robotics platforms, control systems, and automated processes. By augmenting current control and optimization methods, they improve the overall efficiency, flexibility, and resilience of intricate systems and procedures.

3 ABC in Control System

ABC is a search technique based on swarm intelligence that replicates the process of honey bee swarms grazing for food [17]. The ABC algorithm is used for optimizing multivariable functions and results produced by ABC, the Genetic Algorithm (GA), and the Particle Swarm algorithm [13, 36].

3.1 Comparative Study of Various Parameter Use in Control Systems

ABC emulates the foraging behavior of a honeybee swarm. This study employs the ABC algorithm to optimize a vast array of numerical test functions. The outcomes generated by the ABC algorithm are then contrasted with those achieved by the genetic algorithm (GA) and particle swarm optimization algorithm [15]. The objective is to elucidate the Bees technique, an optimization technique that draws inspiration from the foraging activity of honeybees, in order to discover the most advantageous solution. The algorithm conducts both an exploitative search in the vicinity and a random exploratory search [40]. The results indicate that the ABC technique utilizes bee agents to collect food sources and subsequently adjust the fitness function. This function is employed to determine the most ideal test cases with the largest path coverage while minimizing resource usage.

3.2 Optimal Control Strategies

ABC algorithm is an optimization algorithm based on the intelligent behavior of honeybee swarms [7]. It produces successful results in many application studies, resulting in better performance from ABC. There are specific tasks that some individuals carry out in a real bee colony. These specialized bees maximize the nectar amount stored in the hive by performing efficient division of labor and self-organization. The minimal model of swarm- intelligence for age selection in a honeybee colony, on which the ABC algorithm depends, consists of three kinds of bees: employed bees, onlooker bees, and scout bees.

An employed bee travels to the food source it previously visited by itself, while an onlooker bee waits on the dance floor to decide which food source to choose [16]. A bee carrying out a random search is called a scout.

3.3 Adaptive Control

Adaptive control detects changes in the functioning of the process. The core idea is to design an ABC variant that can adjust its parameters or control strategy based on the system's behavior. In this, an employed bee visits a new food source that is near the location of a previously visited one. Employed Bee applies a greedy selection between them and then turns back to sharing information about the memorized food source with the onlooker bees. However, it is not straightforward, and they actually managed the nondeterministic behaviors of the employed bees [3].

3.4 Case Studies in Control System Optimization

Dervis Karaboga et al. presented the results of the performance of the ABC algorithm for constrained optimization problems. In the ABC-PID controller, the ABC algorithm is used to optimize gains. The main objective is to optimize the gains of the PID controller for the plants. A PID controller is widely used for speed and position control [1]. We consider a bee colony as a dynamic system gathering information from an environment and adjusting its behavior in accordance with it [15]. ABC simulates the intelligent foraging behavior of a honeybee swarm. The honeybee colony is chosen among the different nectar sources available in the field. Selectively foraging from those that are most profitable. The ABC algorithm is used to improve a lot of numerical test functions in this work. The outcomes of the ABC algorithm are compared to those of genetic algorithms, particle swarm optimization algorithms, differential evolution algorithms, and evolution strategies. The results show that the performance of the ABC is better than or like that of other population-based algorithms, with the advantage of employing power control parameter size gains. The

Table 1 Discussion on ABC in the control system

Algorithm	Application	Remarks	References
ABC	Path planning	Optimize multivariable functions	[15]
Genetic algorithm	Natural evolution	Explicit exchange of genetic material	[16]
Particle swarm algorithm	Particle swarm optimization	Optimize non-linear and multidimensional problems	[25]

main objective is to optimize the gains of the PID controller for the plants. A PID controller is widely used for speed and position control [1].

A bee colony is seen as a dynamic system that collects information from its surroundings and adapts its behavior accordingly [15]. ABC emulates the intelligent foraging behavior of a swarm of honeybees. The honeybee colony selects among the several nectar sources present in the field. Engaging in selective foraging by targeting the most profitable options. In this study, the ABC algorithm is employed to enhance various numerical test functions. The results of the ABC algorithm are compared to those of genetic algorithms and particle swarm optimization algorithms (Table 1). The results indicate that the performance of the ABC algorithm is superior to or comparable to that of existing population-based algorithms, with the added benefit of utilizing power control factors.

4 ABC in Robotics

There are many meta-heuristic algorithms, but the ABC algorithm [14] is one of the best and most famous. It is based on how a bee colony finds food. As a swarm intelligence program, ABC can find the best answer by starting search processes in many places at the same time [17]. It has been shown that ABC is better than other algorithms because it can explore more and has fewer control factors.

Recent research on ABC algorithms in robotics has demonstrated a range of inventive uses and approaches. Jevtić et al. [12] propose an enhanced version of the Bees Algorithm that enables robotic bees to perform multi-target search and coverage. Their study showcases the potential of these bees to self-organize, operate without central control, and adapt to unfamiliar environments. In this study, Karthik et al. [18] investigate the application of blockchain technology in swarm robotics to facilitate decentralized communication. This technology allows for the development of 2D structures without the need for a central authority [18]. Verlekar and Joshi [34] simulate the behaviors of ants and bees in swarm robots to improve their foraging abilities and energy efficiency. They use computer vision techniques for navigation purposes. The review by Singh [30] examines the current state of ABC optimization in many fields, highlighting its effectiveness and identifying future research directions. Finally, [4] employ ABC programming to solve the artificial ant problem by optimizing robotic path planning. Their approach yields favorable outcomes when compared to genetic programming. Recent applications of ABC and its variants in robotics have been listed in Table 2.

Recent research on ABC algorithms for robot path planning has resulted in notable progress [22]. In their study, Cui et al. [5] created a more advanced version of the multi-objective ABC algorithm called IMOABC. This algorithm demonstrated better performance in path-planning tasks for mobile robots. It achieved this by employing external archive pruning and non-dominated ranking algorithms [5]. The main purpose of this algorithm was to optimize both path length and safety [39]. Ren et al. [27] proposed a novel approach that combines a structure based on parasitism relations with particle swarm optimization (PSO) and ABC algorithms to improve the efficiency and safety of path planning [27]. Khalil [19] introduced a hybrid modified ABC-PSO algorithm to tackle the issue of slow convergence rate in ABC and enhance the performance of path planning [19]. In 2023, Manish et al. introduced an ABC algorithm for planning the motion path of mobile robots. The method was shown to be effective in navigating unfamiliar surroundings [23]. Recent applications of ABC in robotics path planning have been presented in Table 3.

Table 2 ABC in robotics

Algorithm	Application	Remarks	Referencse
Modified bees algorithm	Multi-target search	Demonstrates a modified Bees Algorithm for self-organized, decentralized multi-target search and coverage in unknown areas using robotic bees	[12]
Blockchain-based decentralized swarm	Decentralized construction	Applies blockchain for decentralized communication in swarm robotics, enabling construction tasks without a central authority	[18]
Ant and bee inspired foraging	Foraging efficiency	Emulates ant and bee behaviors for efficient foraging and energy usage in swarm robots, employing computer vision for navigation	[34]
ABC optimization review	Optimization review	Reviews artificial bee colony optimization techniques and their applications in complex engineering problems, highlighting future research directions	[30]
ABC programming for robotic path planning	Path planning	Uses ABC programming for solving robotic path planning in the artificial ant problem, showing comparative performance against genetic programming	[4]

Table 3 ABC in Robotics Path Planning

Algorithm	Application	Remarks	References
ABC	Path planning	Optimize fitness function to avoid obstacles in dynamic path planning for mobile robot	[22]
IMOABC	Path planning	Proposes IMOABC, demonstrating superior performance in complex multi-objective optimization problems for mobile robot path planning	[5]
IMO-ABC	Path planning	Introduces IMO-ABC for optimizing path length and safety in multi-objective path planning, verified through numerous simulations and statistical analyses	[39]
ABC	Path planning	Combines PSO and ABC in a parasitism-relation-based structure, enhancing path planning safety and cost efficiency	[27]
MABC-PSO	Path planning	Proposes MABC-PSO, improving path planning performance by addressing ABC's slow convergence and poor exploitation	[19]
ABC	Path planning	Introduces an ABC algorithm for mobile robot navigation in unknown environments, demonstrating its effectiveness using MATLAB simulations	[23]

5 ABC in Automation

Automation is one of the popular technologies used in various applications, which reduces the involvement of manual efforts [8] and handles multi-objective optimization tasks by adaptively adjusting parameters, which accelerates this mechanism using ABC algorithms. Automation has a greater level of integration with AI, robotics [9], and other technologies, where techniques like ABC algorithms give an optimized solution to automated systems. In automation, using ABC increases productivity, optimizes the process, and makes it more efficient for automated systems.

5.1 Process Optimization and Various Scheduling Algorithms

ABC algorithms are optimization algorithms to get global optima and various modifications have been done over time which continues to result in guided solutions using the current solution [29], which effectively gives the optimized solution for the automated systems. Effective scheduling in its operations leads to improvement in efficiency [38] and utilization of automated mechanisms. Strategies like task scheduling and load balancing provisioning to optimize the process with system or application requirements.

Tan et al. [32] proposed that Modified ABC using crew scheduling and concluded that MABC has better results than PSO and GA where the mean is reduced by 20%

compared with base time [32]. These scheduling techniques with ABC pertaining to minimizing makespan and maximizing resource consumption in automated systems.

5.2 Energy Efficient and Resource Allocation Method

The optimization of energy in automated systems makes it easier to reach a performance level, and it accomplishes the building of automation. Wahid et al. proposed that to achieve optimum utilization of energy by giving optimized parameters to the fuzzy controller, it is best to maintain efficient energy in automated systems.

Wedde et al. [37] proposed that ABC algorithms inspired by BeeAdHoc initiate a reactive source routing algorithm and consume less energy as it reduces overhead and conserves energy because it utilizes fewer control packets to do routing [37], and this enhances the efficiency of automation by getting leads slow convergence and optimum solution using ABC algorithms inspired by the multi-exchange neighborhood [6], and the exploration capacity of the ABC is constrained in a zone that reduces the computational complexity [33].

5.3 Parameter Optimization in Automated Systems

In an automated system, parameters like convergence, delay, throughput, and behavior are important factors to optimize the ABC algorithms and make them more efficient. The problems that arise due to such parameters (like delay and throughput) require an optimized ABC trial to utilize the food source and explore the routes [10].

Habbi et al. [9] proposed a method in which ABC encoded with a fuzzy model evolved together to optimize a strategy in automation that simulates the global depicting behavior of honey bee swarms [9], and to optimize the parameters (like delay and throughput), they used the routing network AdHoc, which was inspired by the ABC algorithms proposed by Wedde et al. [37], which resulted in the best solution and gave robust automated systems.

5.4 Case Study in Automation

Savsani and Jhala [28] proposed an optimal solution for robot arms that represents the adaptability of a complex environment and scalability to workspaces using the ABC algorithm, which categorizes bees into three groups: employed bees, onlookers, and scouts in 9:11:12, respectively [28].

Case Scenario: The robots are assigned tasks such as picking and placing objects within the warehouse environment. The main two objectives are to avoid collisions with obstacles and minimize the energy consumption of the ABC algorithm.

Table 4 Applications of ABC in Automation

Algorithm	Application	Remarks	References
BEE AdHoc	Routing network	To perform routing and utilize energy	[37]
MNABC	Path optimizing	To perform routing and utilize energy	[6]
ABC (fuzzy system)	Finding structure	Enhanced parameter	[9]
AABC	Scheduling technique	Improve efficiency to optimal solution	[26]
ABC	Process optimization	Optimizing the process of search strategy	[29]

However, its effectiveness depends on the availability of resources and the tuning of parameters.

Their future enhancement would be in the workspace, where dynamic obstacles or real-time adaptation are also required to extend the use of the ABC algorithm in robotics and automation. This approach utilized the ABC algorithm for optimal trajectory and energy consumption of the robot's arm in a free and obstacle-free workspace. Table 4 presents recent applications of ABC for various applications in Automation.

6 Challenge and Conclusion

The scope and aims of implementing ABC algorithms in control systems, robotics, and automation encompass a broad range of optimization tasks targeted at enhancing system performance, efficiency, adaptability, and dependability across numerous domains and applications. These algorithms contribute to the development of intelligent and autonomous systems in various real-world contexts by providing efficient answers to challenging optimization problems. There are a variety of obstacles and factors to consider when using ABC algorithms in robotics, automation, and control systems, as discussed in the following section.

Algorithm Parameter Tuning

Despite their ease of use, ABC algorithms can exhibit significant variations in performance based on the parameters set. This includes the number of bees, population size, trial limit, and probabilities for various bee species (employed, onlooker, and scout). It can be difficult to determine the ideal parameter values for various applications and problem domains; this sometimes necessitates trial and error.

Convergence Speed and Premature Convergence

ABC algorithms can converge to unsatisfactory solutions before covering the whole search space, a condition known as premature convergence. To avoid premature convergence and guarantee that the algorithm sufficiently examines attractive portions of the search space, it is imperative to strike a balance between exploration and exploitation.

Managing Dynamic Environments

The surroundings in robotics, automation, and control systems are often dynamic and susceptible to changes throughout time. To monitor shifts in the problem domain, ABC algorithms need to continuously update their population and exploration methods to effectively adjust to changing circumstances. In order to achieve strong performance in changing situations, it is necessary to create adaptive mechanisms that allow the algorithm to adjust its parameters and behavior as needed.

Complexity and Scalability

ABC algorithms may face substantial difficulties due to the intricacy and scalability of optimization problems in robotics, automation, and control systems. The algorithm's performance may deteriorate as the problem becomes more complex or as the search space becomes more dimensional since it becomes more difficult to explore the whole solution space and more computationally demanding. The scalability of ABC algorithms for large-scale optimization problems can be enhanced by creating effective data structures, optimization methods, and parallelization schemes.

Managing Constraints and Uncertainties

Constraints and uncertainties are common in control systems, robotics, and automation optimization challenges, and they must be taken into account throughout the optimization process. In order to guarantee that solutions are workable and resilient in practical situations, ABC algorithms need to effectively manage constraints and integrate techniques for modeling uncertainty. Robust optimization methods, constraint management systems, and stochastic search approaches can improve the algorithm's capacity to manage uncertainties and restrictions.

Integration with Current Systems and Algorithms

Compatibility, interoperability, and interface design must be carefully taken into account when integrating ABC algorithms with current robotics platforms, control systems, and automation procedures. For smooth operation and compatibility with the entire system architecture, seamless integration with the current algorithms, sensors, actuators, and communication protocols is necessary.

A thorough understanding of the underlying optimization problem, the design principles of the algorithm, and the real-world limitations of the intended application area are necessary to effectively address these issues. ABC algorithms can provide useful optimization capabilities for enhancing the performance, efficiency, and flexibility of robotics, automation, and control.

References

1. Aggrawal A, Mishra AK, Zeeshan A (2014) Speed control of DC motor using particle swarm optimization technique by PSO tunned PID and FOPID. Int J Eng Trends Technol 16(2):72–79

2. Ait Sahed O, Kara K, Benyoucef A, Hadjili ML (2016) An efficient artificial bee colony algorithm with application to nonlinear predictive control. Int J Gen Syst 45(4):393–417. https://doi.org/10.1080/03081079.2015.1086344
3. Aslan S (2019) A transition control mechanism for artificial bee colony (ABC) algorithm. Comput Intell Neurosci 2019:1–24. https://doi.org/10.1155/2019/5012313
4. Boudardara F, Gorkemli B (2018) Application of artificial bee colony programming to two trails of the artificial ant problem. In: 2018 2nd international symposium on multidisciplinary studies and innovative technologies (ISMSIT), pp 1–6. https://doi.org/10.1109/ISMSIT.2018.8567048
5. Cui Q, Liu P, Du H, Wang H, Ma X (2023) Improved multi-objective artificial bee colony algorithm-based path planning for mobile robots. Front Neurorobot 17:1196683
6. Dongli Z, Xinping G, Yinggan T, Yong T (2012) An artificial bee colony optimization algorithm based on multi-exchange neighborhood. In: 2012 Fourth international conference on computational and information sciences, pp 211–214. https://ieeexplore.ieee.org/abstract/document/6300440/
7. Elsisi M, Soliman M, Aboelela MAS, Mansour W (2015) ABC based design of PID controller for two area load frequency control with nonlinearities. TELKOMNIKA Indones J Electric Eng 16(1):58–64
8. Elkhateeb NA, Badr RI (2017) Novel PID tracking controller for 2DOF robotic manipulator system based on artificial bee colony algorithm. Electr Control Commun Eng 13(1):55–62. https://doi.org/10.1515/ecce-2017-0008
9. Habbi H, Boudouaoui Y, Karaboga D, Ozturk C (2015) Self-generated fuzzy systems design using artificial bee colony optimization. Inf Sci 295:145–159
10. Halim HZA, Moin NH (2022) Balance of exploration and exploitation in artificial bee colony for multi products inventory routing problem. Appl Math Sci 16(9):425–434
11. Huang H-C, Chuang C-C (2020) Artificial bee colony optimization algorithm incorporated with fuzzy theory for real-time machine learning control of articulated robotic manipulators. IEEE Access 8:192481–192492
12. Jevtić A, Gazi P, Andina D, Jamshidi MO (2010) Building a swarm of robotic bees. In: 2010 World automation congress, pp 1–6. https://ieeexplore.ieee.org/abstract/document/5665704/
13. Juneja M, Nagar SK (2016) Particle swarm optimization algorithm and its parameters: a review. In: 2016 International conference on control, computing, communication and materials (ICCCCM), pp 1–5. https://ieeexplore.ieee.org/abstract/document/7918233/
14. Karaboga D (2005) An idea based on honey bee swarm for numerical optimization. Technical report-tr06, Erciyes University, engineering faculty, computer. http://abc.erciyes.edu.tr/pub/tr06_2005.pdf
15. Karaboga D, Akay B (2009) A comparative study of artificial bee colony algorithm. Appl Math Comput 214(1):108–132
16. Karaboga D, Basturk B (2007) A powerful and efficient algorithm for numerical function optimization: artificial bee colony (ABC) algorithm. J Global Optim 39(3):459–471. https://doi.org/10.1007/s10898-007-9149-x
17. Karaboga D, Gorkemli B, Ozturk C, Karaboga N (2014) A comprehensive survey: artificial bee colony (ABC) algorithm and applications. Artif Intell Rev 42(1):21–57. https://doi.org/10.1007/s10462-012-9328-0
18. Karthik S, Chandhar NP, Akil M, Chander S, Amogh J, Aditya R (2020) Bee-bots: a blockchain based decentralised swarm robotic system. In: 2020 6th International conference on control, automation and robotics (ICCAR), pp 145–150. https://ieeexplore.ieee.org/abstract/document/9108053/
19. Khalil NI, Abdullah HN, Hassnawi LA (2023) A hybrid modified ABC-PSO algorithm for optimal robotic path planner. In: 2023 15th International conference on developments in esystems engineering (DeSE), pp 27–32. https://doi.org/10.1109/DeSE58274.2023.10100021
20. Kumar A, Kumar V (2017) Artificial bee colony-based design of the interval type-2 fuzzy PID controller for robot manipulator. In: TENCON 2017-2017 IEEE Region 10 conference, pp 602–607. https://doi.org/10.1109/TENCON.2017.8227933

21. Liang J-H, Lee C-H (2015) Efficient collision-free path-planning of multiple mobile robots system using efficient artificial bee colony algorithm. Adv Eng Softw 79:47–56
22. Ma Q, Lei X (2010) Dynamic path planning of mobile robots based on ABC algorithm. In: Wang FL, Deng H, Gao Y, Lei J (eds) Artificial intelligence and computational intelligence, vol 6320. Springer, Berlin, Heidelberg, pp 267–274. https://doi.org/10.1007/978-3-642-16527-6_34
23. Manish, Kumar S (2023) Robot motion path planning using artificial bee colony algorithm. In: Bhateja V, Yang X-S, Chun-Wei Lin J, Das R (eds) Intelligent data engineering and analytics, vol 327. Springer Nature Singapore, pp 555–565. https://doi.org/10.1007/978-981-19-7524-0_49
24. Nayyar A, Nguyen NG, Kumari R, Kumar S (2020) Robot path planning using modified artificial bee colony algorithm. In Satapathy SC, Bhateja V, Nguyen BL, Nguyen NG, Le D-N (eds) Frontiers in intelligent computing: theory and applications, vol 1014. Springer Singapore, pp 25–36. https://doi.org/10.1007/978-981-13-9920-6_3
25. Pereira G (2011) Particle swarm optimization. INESCID Inst. Super. Techno Porto Salvo Port. https://citeseerx.ist.psu.edu/document?repid=rep1&type=pdf&doi=b7be99e3af4beab734965dd5af6b411a9e291d2a
26. Rekaby A, Youssif AA, Eldin AS (2013) Introducing adaptive artificial bee colony algorithm and using it in solving traveling salesman problem. In: 2013 Science and information conference, pp 502–506. https://ieeexplore.ieee.org/abstract/document/6661785/
27. Ren H, Gao L, Shen X, Li M, Jiang W (2023) A novel swarm intelligence algorithm with a parasitism-relation-based structure for mobile robot path planning. Sensors 23(4):1751. https://doi.org/10.3390/s23041751
28. Savsani PV, Jhala RL (2012) Optimal motion planning for a robot arm by using artificial bee colony (ABC) algorithm. Int J Modern Eng Res (IJMER) 2(6):4434–4438
29. Saxena S, Sharma K, Shiwani S, Sharma H (2014) Lbest artificial bee colony using structured swarm. In: 2014 IEEE international advance computing conference (IACC), pp 1354–1360. https://ieeexplore.ieee.org/abstract/document/6779524/
30. Singh A (2022) Social structure to artificial implementation: honeybees: depth and breadth of artificial bee colony optimization. In: Saraswat M, Sharma H, Balachandran K, Kim JH, Bansal JC (eds) Congress on intelligent systems, vol 111. Springer Nature Singapore, pp 271–285. https://doi.org/10.1007/978-981-16-9113-3_21
31. Szczepanski R, Erwinski K, Tejer M, Bereit A, Tarczewski T (2022) Optimal scheduling for palletizing task using robotic arm and artificial bee colony algorithm. Eng Appl Artif Intell 113:104976
32. Tan Y, Shi Y, Tuba M (2020) Advances in Swarm Intelligence: 11th international conference, ICSI 2020, Belgrade, Serbia, 14–20 July 2020, Proceedings, vol. 12145. Springer Nature
33. TSai P-W, Pan J-S, Liao B-Y, Chu S-C (2009) Enhanced artificial bee colony optimization. Int J Innov Comput Inf Control 5(12):5081–5092
34. Verlekar H, Joshi K (2017) Ant & bee inspired foraging swarm robots using computer vision. In: 2017 International conference on electrical, electronics, communication, computer, and optimization techniques (ICEECCOT), pp 191–195. https://doi.org/10.1109/ICEECCOT.2017.8284663
35. Villegas JM, Caraveo C, Mejía DA, Rodríguez JL, Vega Y, Cervantes L, Medina-Santiago A (2021) Intelligent search of values for a controller using the artificial bee colony algorithm to control the velocity of displacement of a robot. Algorithms 14(9):273. https://doi.org/10.3390/a14090273
36. Wang H, Du H, Cui Q, Song H (2022) Artificial bee colony algorithm based PID controller for steel stripe deviation control system. Sci Prog 105(1):003685042210751. https://doi.org/10.1177/00368504221075188
37. Wedde HF, Farooq M, Pannenbaecker T, Vogel B, Mueller C, Meth J, Jeruschkat R (2005) BeeAdHoc: an energy efficient routing algorithm for mobile ad hoc networks inspired by bee behavior. In: Proceedings of the 7th annual conference on genetic and evolutionary computation, pp 153–160. https://doi.org/10.1145/1068009.1068034

38. Wong LP, Puan CY, Low MYH, Wong YW, Chong CS (2010) Bee colony optimisation algorithm with big valley landscape exploitation for job shop scheduling problems. Int J Bio-Inspired Comput 2(2):85. https://doi.org/10.1504/IJBIC.2010.032125
39. Yu Z, Duan P, Meng L, Han Y, Ye F (2022) Multi-objective path planning for mobile robot with an improved artificial bee colony algorithm. Math Biosci Eng 20(2):2501–2529. https://doi.org/10.3934/mbe.2023117
40. Yuce B, Packianather MS, Mastrocinque E, Pham DT, Lambiase A (2013) Honey bees inspired optimization method: the bees algorithm. Insects 4(4):646–662

Integrating Artificial Bee Colony Algorithms for Deep Learning Model Optimization: A Comprehensive Review

Faiz Akram, Shafaque Aziz, Nayyar Ahmed Khan, Syed Akramah Faizi, and Khalid Raza

1 Introduction

Optimization is the process of finding the best possible solution to a problem within a defined set of constraints and criteria which involve adjusting the values of decision variables to maximize or minimize an objective function. Examples of optimization problems include minimizing production costs, maximizing profits, finding the shortest path in a network, and training machine learning models by minimizing error functions [1, 2]. The key elements of optimization include Objective Function, Decision Variables, and Constraints. Optimization problems are ubiquitous in various fields including engineering, economics, finance, logistics, and machine learning [3].

Nature-inspired optimization algorithms draw inspiration from natural processes and phenomena to solve complex optimization problems. These algorithms mimic the behavior of biological, physical, or social systems. The main categories include Evolutionary Algorithms (EAs), Swarm intelligence (SI), Artificial Immune Systems (AISs), and other Bio-Inspired Algorithms (BIAs). EAs are inspired by the process of natural evolution which includes Genetic Algorithms (GA), Genetic Programming

F. Akram · S. Aziz
Department of Computer Engineering, Faculty of Engineering and Technology, Jamia Millia Islamia, New Delhi, India

S. Aziz
Government Engineering College Samastipur, Bihar, India

N. A. Khan
Department of Computer Science, College of Computing and Information Technology, Shaqra University, Shaqra, Saudi Arabia

S. A. Faizi
Department of Information Technology, Delhi Technological University, New Delhi, India

K. Raza (✉)
Department of Computer Science, Jamia Millia Islamia, New Delhi, India
e-mail: kraza@jmi.ac.in

K. Raza (ed.), *Solving with Bees*, Springer Tracts in Nature-Inspired Computing,
https://doi.org/10.1007/978-981-97-7344-2_5

(GP), and Differential Evolution (DE) [4]. They use mechanisms such as selection, mutation, and crossover to evolve solutions over generations. SI is inspired by the collective behavior of decentralized, self-organized systems, particularly the behavior of animals that live in colonies or swarms. Examples include Ant Colony Optimization (ACO) [5] and Particle Swarm Optimization (PSO). AIS is based on the human immune system's ability to recognize and adapt to new threats. These algorithms use concepts like immune memory and clonal selection. BIAs include a variety of algorithms inspired by different natural processes, such as the Artificial Bee Colony (ABC) algorithm (mimicking the foraging behavior of honey bees), Firefly Algorithm (based on the flashing behavior of fireflies), and Cuckoo Search (inspired by the brood parasitism of some cuckoo species), and so on. These algorithms are particularly powerful in solving optimization problems that are too complex for traditional analytical methods, providing approximate solutions efficiently [6].

The ABC algorithm is a popular nature-inspired optimization technique based on the foraging behavior of honey bees. It has multiple roles in deep learning including *Feature Selection, Model Structure Optimization, Hyperparameter Optimization*, and *Training Neural Networks* (Fig. 1).

Hive learning is a concept that takes inspiration from the collective intelligence exhibited by social insects like bees, ants, and termites. It focuses on distributed learning and collective problem-solving, where multiple agents (or "hives") work together to achieve a common goal [7]. In the context of machine learning and deep learning, hive learning can be interpreted as *Distributed Learning, Ensemble Methods, Collaborative Learning*, and *Swarm Intelligence*. In distributed learning, multiple learning agents (e.g., machines or processors) work collaboratively to train a model or solve a problem. This approach is particularly useful for large-scale data and complex models that require significant computational resources. Hive learning is akin to ensemble learning methods where multiple models (learners) are combined to improve overall performance. Techniques like bagging, boosting, and stacking leverage the collective intelligence of several models. In environments like federated learning, multiple devices or organizations collaborate to train a shared model without exchanging raw data, preserving privacy and improving robustness. Hive learning utilizes principles from swarm intelligence, where simple agents follow simple rules to exhibit complex behavior collectively. This can be applied to optimize neural network training, parameter tuning, and other aspects of machine learning. In a nutshell, nature-inspired optimization algorithms such as ABC has significant applications in optimizing various aspects of deep learning models [8]. Hive learning represents a powerful paradigm for distributed and collaborative machine learning approaches. Together, these concepts enhance the efficiency, scalability, and performance of artificial intelligence (AI) systems.

Rest of the chapter is organized as follows. Section 2 explains background of ABC algorithms, while Sect. 3 explains deep learning models and optimization challenges. Section 4 showcases the integration of ABC algorithms in deep learning for various optimization problems. Section 5 presents a few case studies and applications; Sect. 6 explains hybridization of ABC with other metaheuristic algorithms.

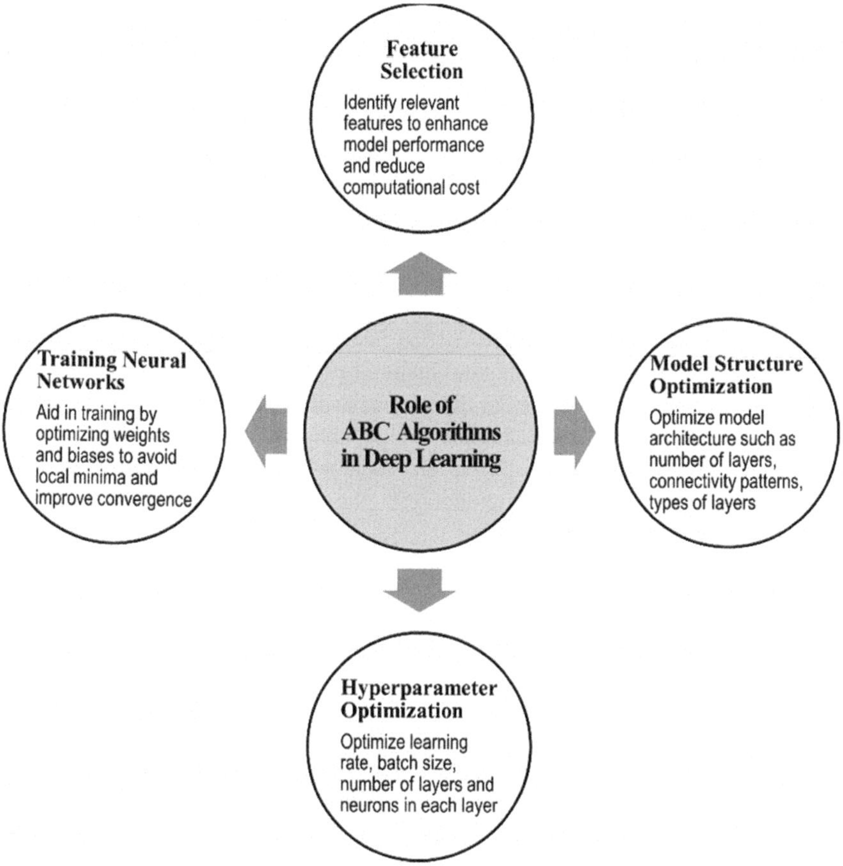

Fig. 1 Role of Artificial Bee Colony (ABC) algorithms in deep learning

Section 7 presents challenges and future research directions in the field, and finally, Sect. 8 concludes the chapter.

2 Background of ABC Algorithms

2.1 Introduction to ABC Algorithms

Several research problems require the optimization of data for proper statistical analysis. Karaboga [9] proposed the artificial bee colony algorithm to optimize numerical problems. The algorithm focuses on simulating the behavior of honey bee swarms. The proposed model comprises employed bees, onlookers, and scouts. The algorithm focuses on the intelligent behavior of foraging bees in the honeybee colonies. The first two employees search for proper food in different locations, whereas the third section remains close to their hive. The proposed model simulates the self-organizing characteristic of the honeybee. It also portrays collective intelligence, thereby simulating the intelligent behavior of the employed bees. The ability to take positive or negative feedback to accept or reject rich food sources is used as intelligence in this approach. The simulation of the ABC algorithm is a reference to find the most optimal solution for a computational problem. In this approach, it is necessary to identify the best parameters first to find the best optimal solution. On placing the parameters, the objective function can be minimized. The algorithm is inspired by foraging behavior, which identifies optimal solutions from the existing vectors. This algorithm is assumed to be a global optimal solution for finding the optimization value from extensive data. Several combinatorial problems are also solved with the help of this algorithm [10]. Several researchers, such as [11, 12], explain the use of ABC optimization for unconstrained as well as constraint optimization problems. An exciting feature of this algorithm refers to the user's employment of only three predetermined values. The total population set, the maximum limit, and the number of iterations are required to control the algorithm.

2.2 *Exploration of Foraging Behavior in Natural Bee Colonies*

The behavior of the bees is identified in three different phases. The initial phase determines the food sources from the existing population of all directors. The solution vector from this phase refers to identifying the most appropriate optimized food source. The analogous value for this food source in the real world is the identification of the vector having a minimal objective function. During the initial phase, the scout bees complete the identification of all the sources. Once the sources are identified, the employed bees and onlooker bees identify the best food sources. Due to continuing working, the onlooker bees get exhausted and require a break. All the employed bees exploiting the food sources were converted to Scout bees, in this case, to search for further sources of good nectar. All employed bees who support sources are exploited entirely and turn into Scout bees. As an analogy, the food sources represent the solution sets for an optimization problem. The quality of the food means the identification of a proper fitness value for the suggested optimal solution. It is worth mentioning that the total number of food sources (probable solutions) equals the total number of employed bees in this case because each employee is associated with one or the other food source.

2.3 ABC Algorithm Components and Workflow

The ABC algorithm executes in four different stages. Initially, the initialization of the vectors from the population of the food sources is done by the scout bees. After setting all the components and control values, initialization takes place between the upper and lower limit as the equation below as adapted from [13]:

$$x_m = l_i + rand(0,1) * (u_i - l_i) \tag{1}$$

where x_m refers to the food sources, u_i and l_i are the upper and lower bounds of the values.

The next phase of the algorithm is to identify new food sources with better-quality nectar. The employed bees try to find the most appropriate solutions from new elements in the population set. A better food source can be determined with the help of the formula created below

$$v_m = x_{mi} + \Phi(x_{mi} - x_{ki}) \tag{2}$$

where v_m refers to the new food sources, x_{mi} and x_{ki} are randomly selected food sources and initial food sources in the population set. The fitness function Φ is a random value between a given range. Greedy approaches are applied after the calculation of the new food source. Better results are accepted and the previously existing results are rejected. In case the previous results are better than the newer ones are rejected.

The onlooker be phase reserves the ability to identify the most appropriate solution from the existing sources. The onlooker bees use the fitness function adapted by the employed bees to predict the probability of calculating acceptable food sources [13].

$$P_i = \frac{fit_i}{\sum fit_j} \tag{3}$$

Once the food sources are accepted based on the probabilistic selection, neighbourhood sources are identified and recruited for their fitness value. A similar technique is adapted as in Eq. (2) above for the newly chosen food source. All the employed bees who are responsible for the selection of food sources are called scout bees. They are responsible for randomly searching for newer and better food sources to enrich the content of the nectar. If any solution is not feasible, the scout bees propose a more unique solution for adaptability. The poor solution with negative values or feedback is rejected in this case. Appropriate results are accepted with the help of the scout's evaluation of the new solutions.

3 Deep Learning Models and Optimization Challenges

3.1 Brief Overview of Deep Learning Models

Recent advancements in deep learning have significantly reshaped the field of system modeling and intelligence. Comparative studies consistently demonstrate the superiority of deep learning and hybrid machine learning models compared to traditional machine learning approaches. Deep learning methods are undergoing rapid development, constantly pushing the boundaries of performance.

A wealth of literature exists, including comprehensive review papers, that delve into the evolution of deep learning algorithms within specific application domains. The integration of ABC algorithms with deep learning has found application across various fields of study, showcasing its versatility and effectiveness in optimizing complex tasks, including energy demand forecasting [14], medical image analysis [15, 16–18], radiology [19], multimedia analytics, sentiment analysis [20], human recognition [21], transportation systems [22, 23], radar signal processing [24], and so on. These studies underscore the broad impact and potential of integrating ABC algorithms with deep learning across diverse fields, addressing complex challenges and advancing capabilities in artificial intelligence and machine learning research.

Some prominent architectures form the core of many successful deep learning applications as follows [25]:

i. **Convolutional Neural Networks (CNNs)**: These networks excel in image processing tasks due to their specialized architecture. They consist of three main types of layers: convolutional layers for feature extraction, pooling layers for down-sampling, and fully connected layers for classification. Training involves two main stages: the feed-forward stage where data travels through the network, and the backpropagation stage where the network adjusts its internal parameters to minimize errors.

ii. **Recurrent Neural Networks (RNNs)**: Designed for sequence and pattern recognition in applications like speech, handwriting, and text analysis, RNNs utilize cyclic connections within their structure. This allows them to process sequential data by considering past information in their current calculations. Unlike traditional neural networks, RNNs have connections that feed information not only to the next layer but also back to the same layer in the next time step. This information is stored in a "state vector" within hidden units and used to compute outputs [26].

iii. **Denoising Autoencoders (DAEs)**: An extension of the autoencoder (AE) architecture, DAEs are designed specifically for learning features from noisy data. They consist of three main layers: an input layer, an encoding layer that compresses the data, and a decoding layer that reconstructs the input from the compressed representation. DAEs can be stacked upon each other to create a **Stacked Denoising Autoencoder (SDAE)**, an unsupervised algorithm used for nonlinear dimensionality reduction in deep learning models. This method employs a feed-forward structure with multiple hidden layers and a pre-training strategy to learn features before fine-tuning them for a specific task.

iv. **Deep Belief Networks (DBNs)**: These networks excel at learning complex relationships in high-dimensional data. They are composed of multiple layers with connections between layers but not within individual layers. DBNs can be considered a hybrid network containing both directed and undirected connections. Each layer in a DBN typically consists of a **Restricted Boltzmann Machine (RBM)**, a building block trained in a greedy layer-wise manner. RBMs contain only two layers: a visible layer representing the input data and a hidden layer that extracts features. DBNs are often praised for their high accuracy and computational efficiency in various applications.

v. **Long Short-Term Memory (LSTM)**: An extension of RNNs, LSTMs address the vanishing gradient problem that can hinder learning in traditional RNNs for long sequences. They incorporate specialized "gates"–input, output, and forget gates–that control the flow of information within the network. These gates allow LSTMs to selectively remember and forget information relevant to the current task, making them highly effective for tasks involving long-term dependencies within sequences. Their applications span beyond sequence recognition and image processing, showcasing great potential in various environmental fields.

3.2 *Challenges in Training Deep Neural Networks*

Training deep neural networks, while powerful, presents several hurdles. Here are some key challenges:

i. **Overfitting**: Generally, increasing the number of units and layers in a network can enhance its expressive power, allowing it to capture more complex relationships in the data. However, this complexity can lead to a more intricate cost function and a higher susceptibility to getting stuck in suboptimal solutions within the training data (local minima). This results in poor generalization performance (overfitting) where the model performs well on the training data but poorly on unseen data [27].

ii. **Vanishing and Exploding Gradients**: Deep learning typically relies on deep neural networks with many layers of interconnected processing units (neurons). While backpropagation, a common training algorithm, remains applicable to these networks, its effectiveness diminishes with increasing network size. Two main challenges arise [27].

- *Vanishing Gradients*: When using small weight values to initialize the network, the gradients calculated during backpropagation become progressively smaller as they travel through numerous layers, ultimately approaching zero due to numerical limitations. This results in negligible updates to the weights in earlier layers, hindering learning.
- *Exploding Gradients*: Conversely, using large initial weight values can lead to the opposite problem, where the gradients become progressively larger during

backpropagation, causing the learning process to diverge and become unstable from the start, preventing convergence to an optimal solution.

iii. **Data Challenges**:
   - *Limited Labelled Data*: Acquiring and labelling large amounts of data can be expensive and time-consuming.
   - *Data Imbalance*: Datasets may have uneven class distributions, affecting learning.
   - *Label Noise*: Mislabelled data can mislead the network, hindering accurate learning.

iv. **Computational Resources**: Training large and complex models requires significant computing power and memory, limiting accessibility.

v. **Hyperparameter Tuning**: Finding the optimal settings for various parameters (learning rate, batch size, etc.) is crucial but can be time-consuming and require expertise.

3.3 The Role of Optimization in Improving Model Performance

Training deep learning models involves adjusting their parameters to minimize a loss function, which measures the model's performance discrepancy from the desired outcome. This process typically occurs in a supervised learning setting, where the model receives both data samples and their corresponding expected outputs. During training, the model compares its generated outputs with the desired outputs and computes the difference (loss). Optimization algorithms then iteratively refine the model's parameters to minimize this loss, gradually improving its accuracy.

There's no one-size-fits-all solution for model optimization, as different problems might require different optimization algorithms. Researchers have actively investigated various methods to find the best approach for each specific case [28].

Here are some commonly used optimization algorithms in deep learning [28].

i. **Stochastic Gradient Descent (SGD):** This algorithm updates the model's parameters based on the error (loss) calculated for individual data points within the training dataset. It's a faster alternative to the standard Gradient Descent, which uses the entire dataset for each update.

ii. **Nesterov Momentum:** This method builds upon the traditional Momentum algorithm by incorporating an estimation of future parameter positions during the update process, leading to faster convergence compared to standard Momentum.

iii. **Adagrad:** This algorithm adapts the learning rate for each parameter individually based on its historical squared gradients. Parameters with larger historical gradients receive lower learning rates, while those with smaller gradients experience higher learning rates. This helps prevent parameters with frequent updates from dominating the learning process.

iv. **Adadelta**: This algorithm is an extension of Adagrad that addresses its potential accumulation of excessive gradients over time. Instead of accumulating all historical gradients, Adadelta uses a fixed-size window to track only the recent gradients, improving stability and convergence.
v. **RMSProp (Root Mean Square Prop)**: Similar to Adagrad, RMSProp adapts the learning rate for each parameter based on its historical gradients. However, it uses an exponentially decaying average of the squared gradients instead of directly accumulating them, leading to smoother learning updates and faster convergence compared to Adagrad.
vi. **Adam (Adaptive Moment Estimation)**: This algorithm combines the strengths of both RMSProp and Momentum. It utilizes an exponentially decaying average of both gradients and squared gradients, along with a bias correction mechanism, to adaptively adjust learning rates and achieve faster convergence.

4 Integration of ABC Algorithms in Deep Learning

4.1 Rationale for Combining ABC Algorithms with Deep Learning

Combining ABC algorithms with deep learning offers several advantages, enhancing the performance and efficiency of deep learning models. Some of the main rationales for this combination is described as follows [29, 30]:

Feature Selection: Deep learning models, especially those dealing with large datasets, can benefit from effective feature selection and extraction. ABC algorithms can be employed to select the most relevant features, reducing the dimensionality of the data and improving the model's generalization capabilities. This can lead to faster training and better performance by focusing on the most informative features.

Hyperparameter Optimization: Deep learning models have numerous hyperparameters (e.g., learning rate, batch size, number of layers, number of neurons) that significantly impact their performance. Traditional methods for hyperparameter tuning, like grid search or random search, can be inefficient and time-consuming. ABC algorithms, with their ability to explore and exploit the search space efficiently, can effectively optimize these hyperparameters, leading to better model performance and reduced training time.

Optimizing Model Architecture: Designing the architecture of deep learning models (e.g., the number of layers, types of layers, connectivity patterns) is a complex task often done manually or through trial and error. ABC algorithms can automate this process by optimizing the model architecture, leading to more efficient and potentially novel configurations that might be difficult to design manually.

Avoiding Local Minima and Improved Convergence: Deep learning models, particularly those with many layers and complex structures, often encounter local minima during training, which can prevent the model from reaching the optimal solution. ABC algorithms are designed to escape local optima by using a combination of exploration (searching new areas) and exploitation (refining known good solutions). Traditional gradient-based optimization methods, like stochastic gradient descent, can sometimes struggle with convergence, especially in highly non-convex landscapes typical of deep learning problems. ABC algorithms, with their robust search capabilities, can help improve convergence rates by providing diverse and high-quality initial solutions, which can guide gradient-based methods more effectively.

Training Efficiency: ABC algorithms can be integrated into the training process of deep learning models to optimize the weights and biases. This can complement traditional backpropagation methods, potentially leading to faster convergence and better performance by avoiding poor local minima and finding more optimal weight configurations.

Hence, combining ABC algorithms with deep learning leverages the strengths of both approaches. ABC algorithms provide robust optimization capabilities, which can enhance the hyperparameter tuning, feature selection, and overall training process of deep learning models. This synergy can result in models that are not only more accurate and efficient but also more robust to the complexities inherent in deep learning tasks.

4.2 Techniques for Integrating ABC into the Optimization Process

Integrating ABC algorithms into the deep learning optimization process can be done through several techniques. These techniques leverage the strengths of ABC for various stages of deep learning, such as hyperparameter tuning, weight optimization, and feature selection. For *Hyperparameter Optimization*, ABC can be used to optimize hyperparameters by treating the selection of hyperparameters as an optimization problem. It can be done by first defining an objective function that measures the performance of the deep learning model (e.g., validation accuracy, loss), then defining the search space for hyperparameters such as learning rate, batch size, number of layers, and number of neurons, and finally using the ABC algorithm to explore and exploit the search space, evaluating the objective function for different sets of hyperparameters and iteratively improving the solutions. Similarly, for *Weight Initialization and Optimization*, ABC can be integrated into the weight initialization or optimization phase to improve the training process. Firstly, ABC can be used to generate an initial set of weights for the deep learning model, which can then be refined using traditional gradient-based methods, followed by combining ABC with gradient descent. For example, use ABC to periodically adjust weights during training to avoid local minima and guide the gradient descent process. For *Feature Selection*, ABC can be employed for feature subset selection by representing each bee as a subset of features. The fitness of each bee can be evaluated based on the model's performance using that subset. Further, ABC algorithm can be utilized to iteratively refine the feature subset, improving the quality of the selected features [29].

For *Architecture Optimization*, represent each bee as a candidate network architecture, including the number of layers, types of layers, and connectivity patterns. Further, evaluate each architecture based on model performance metrics (e.g., accuracy, loss) and use ABC to explore and refine the architecture design. For *Learning Rate Scheduling*, ABC can optimize learning rate schedules dynamically during the training process: Use ABC to adjust the learning rate at different stages of training, aiming to improve convergence and avoid issues like vanishing or exploding gradients [30].

4.3 Comparative Analysis with Other Optimization Algorithms

The ABC algorithm is a nature-inspired metaheuristic optimization technique designed to mimic the foraging behavior of honey bees. It has gained traction in deep learning due to its simplicity, flexibility, and promising results across various tasks. However, it's crucial to compare it to other popular optimization algorithms to understand its strengths and weaknesses in the context of deep learning. A comparative analysis of ABC with other commonly used algorithms is presented in Table 1.

ABC offers a compelling alternative to traditional optimization algorithms for deep learning. While its simplicity and exploration–exploitation balance are valuable

Table 1 Comparison of ABC with other commonly used optimization algorithms

Algorithm	Strengths	Weaknesses	Deep learning applications
Artificial Bee Colony (ABC)	Simple implementation	Slower convergence in high-dimensional problems	Hyperparameter tuning
Stochastic Gradient Descent (SGD)	Fast convergence	Can get stuck in local optima	Widely used in various deep learning tasks
Adam	Adaptive learning rates	Can be computationally expensive	Often outperforms SGD in complex tasks
RMSprop	Robust to sparse gradients	May not converge well with noisy data	Useful for recurrent neural networks (RNNs)
Particle Swarm Optimization (PSO)	Efficient exploration of search space	Prone to premature convergence	Feature selection

features, its slower convergence in high dimensions and dependence on memory limit its applicability in certain scenarios. Continuous research and hybrid approaches hold promise for further improving ABC's effectiveness in deep learning optimization.

4.4 Optimal Parameter Settings for ABC in Deep Learning

Unfortunately, there is no single "optimal" parameter setting for the ABC algorithm across all deep learning applications. The optimal configuration depends heavily on several factors, including [31].

i. Deep Learning Model and Task:
- **Network Architecture:** Different network architectures, like CNNs, RNNs, or transformers, may require different parameter settings for ABC to navigate the complex optimization landscape effectively.
- **Task Type:** Image classification, machine translation, or time series forecasting might benefit from different ABC parameter settings due to their inherent characteristics and data distributions.

ii. Dataset Properties:
- **Size and Complexity:** Larger and more complex datasets might necessitate adjustments in ABC parameters, like the number of employed bees or iterations, to achieve sufficient exploration and exploitation.
- **Data Characteristics:** The presence of noise, imbalance, or outliers in the data could influence the choice of ABC parameters like learning rate or stopping criteria.

iii. Computational Resources:
- **Available Hardware:** The computational power of your hardware (CPU, GPU) can influence settings like population size and maximum iterations. Running the algorithm on a machine with limited resources might require smaller populations or fewer iterations, impacting performance.

General Guidelines (Not Optimal Values)

While there's no universally optimal setting, here are some commonly used ranges for ABC parameters in deep learning applications:

- Number of employed bees (employed_bees): 10 - 50
- Number of onlooker bees (onlooker_bees): Employed bees * 2
- Number of scout bees (scout_bees): Employed bees / 2
- Maximum iterations (max_iter): 100 - 500
- Learning rate (learning_rate): 0.1 - 1.0

Approaches for Finding Optimal Settings

- **Grid Search or Random Search:** Systematically exploring different parameter combinations can help identify a good starting point or region for further refinement.

- **Adaptive Parameter Tuning:** Techniques like Bayesian optimization or evolutionary algorithms can learn from previous trials and adjust parameters dynamically, potentially leading to an improved configuration.
- **Leveraging Experience and Research:** Consulting existing research papers and studies using ABC for similar tasks can provide valuable insights into effective parameter settings.

4.5 Sensitivity Analysis to Understand Algorithm *Behavior*

Sensitivity analysis serves as a valuable instrument in the realm of ABC for comprehending the behaviour and performance of algorithms. It provides researchers with insights into how optimization outcomes are influenced by algorithm parameters and design decisions. The application of sensitivity analysis to ABC algorithms is as follows:

i. **Parameter Sensitivity Analysis**: An analysis of the way the performance of the ABC algorithm is affected by variations in the method's parameters, including neighborhood size, exploration–exploitation balance, population size, and convergence criteria. Through meticulous manipulation of these factors and observation of their impact on optimization results, researchers can pinpoint crucial parameters that greatly impact the performance and behavior of the algorithm. We compute partial derivatives of each metric like convergence speed, quality of solution, etc. with respect to parameters Q of ABC algorithm and rank them based on the magnitude of each derivative [32].
ii. **Initial Solution Sensitivity**: The initial sensitivity refers to the sensitivity of the ABC algorithms to initial populations that are generated randomly or to beginning solution configurations [33]. Through a comparison of optimization results obtained with different beginning solutions, the robustness of ABC is evaluated by algorithms and scenarios where the choice of starting solution may considerably affect convergence behavior, and solution quality is identified. The sensitivity is calculated by getting the variance of the initial parameters.
iii. **Neighborhood Structure Sensitivity**: The ABC algorithms' susceptibility to changes in search space representations and neighborhood patterns. It investigates how various neighborhood architectures, such as topological or Euclidean distance-based neighborhoods, can affect the ABC algorithms' capacity for exploration and exploitation [34]. This sensitivity analysis is a useful tool for determining which neighborhood topologies work best for various optimization objectives. The ABC algorithm's neighborhood structure is determined mathematically, detailing the generation and evaluation of candidate solutions inside the search space.

iv. **Algorithmic Variants Sensitivity**: The comparative sensitivity of various versions or adaptations of ABC algorithms to alterations in algorithmic parameters and design choices is assessed. For instance, the effectiveness of adaptive strategies, local search mechanisms, and hybridization with other optimization approaches are compared with the typical ABC algorithms [35]. Sensitivity analysis can help identify the algorithmic variants that perform best in various optimization settings.
v. **Problem Instance Sensitivity:**
vi. The ABC algorithms are highly sensitive to variations in issue instances, such as changes in problem dimensionality, complexity, and structure. By applying ABC algorithms to a wide range of benchmark problems that possess varied characteristics, experts can evaluate the algorithm's capacity to adjust to various optimization landscapes and pinpoint problem instances where the algorithm may encounter difficulties or demonstrate exceptional performance [36].

4.6 Guidelines for Choosing Appropriate Parameters for Specific Tasks

Optimal parameter selection in the ABC algorithm necessitates meticulous evaluation of multiple criteria, such as issue attributes, optimization objectives, computational capabilities, and algorithmic limitations [37]. There are various guidelines to choose appropriate parameters for any specific task which are as follows:

i. **Understanding the Problem Characteristics**: The features of the optimization problem such as the objective function, dimensionality, complexity, and constraints need to be analyzed [38]. It is important to select parameters that fit the dynamics and structure of the specific problem.
ii. **Examining Sensitivity Analysis**: Sensitivity analysis must be evaluated to comprehend the impact of parameter values on convergence behavior, solution quality, and computing efficiency once default parameter values are considered, which establishes a baseline for the ABC method [36].
iii. **Adjusting Exploration–Exploitation of the Algorithm**: Optimizing factors associated with exploration and exploitation tactics, including the number of scout bees, hired bees, and observer bees [39]. Strive to explore various regions of the search space while also focusing on refining promising solutions.
iv. **Modifying the dimensions of the Neighborhood**: It is possible to adjust the neighborhood size parameter, which controls the proximity of potential solutions that are investigated by the algorithm. Neighborhoods with larger sizes encourage global exploration, whereas neighborhoods with smaller sizes make it easier to exploit the local community [40].

v. **Population Size**: A suitable population size is chosen to strike a balance between computing efficiency and exploration capabilities. While larger populations allow for more varied exploration, they may also come with a higher computing cost. Try out various population sizes to determine the right balance [41].
vi. **Specifying the criteria for convergence**: The definition of convergence criteria is based on the requirements of the optimization work and the limitations of computational resources [42]. The termination condition of the algorithm is influenced by parameters such as maximum iterations, convergence threshold, or stagnation criteria.

5 Case Studies and Applications

5.1 Real-World Examples of ABC Integration in Deep Learning Tasks

The ABC method has been effectively incorporated into deep learning tasks in several real-world applications. The following examples address a variety of applications that are found in the real world:

5.1.1 Medical Image Analysis

Classification and segmentation of medical images, including MRI and CT scans, are performed. By optimizing the hyperparameters of CNNs, ABC enhances the efficacy of these networks in anomaly detection and classification. ABC is employed to enhance the efficiency of CNNs in identifying and categorizing anomalies by optimizing their hyperparameters. It is possible to identify diseases like tumors or lesions with greater accuracy, which helps with early diagnosis and therapy planning [43].

5.1.2 Text Classification

The ABC algorithm is utilized to optimize the weights and topologies of neural networks, such as RNNs or transformers, in text classification applications such as sentiment analysis and spam detection [44]. This integration improves the model's capacity to analyze and classify textual input, resulting in more efficient spam filters and sentiment analysis tools [45].

5.1.3 Intrusion Detection Prediction

ABC is employed in the field of cybersecurity to enhance the performance of deep learning models in Intrusion Detection Systems (IDS) [46]. This enhances the system's capacity to promptly identify and react to malevolent behaviors and unauthorized entry, consequently augmenting the detection rates of cyber threats and improving network security.

5.1.4 Stock Market Prediction

ABC utilizes fine-tuning techniques to optimize the parameters of deep learning models, specifically LSTM networks, to enhance the precision of stock price and market trend predictions in the field of stock market forecasting [47]. This aids traders and investors in making well-informed selections based on dependable predictions.

5.1.5 Autonomous Navigation

ABC enhances deep reinforcement learning algorithms to optimize autonomous navigation in robotics, resulting in enhanced efficiency and safety in path planning and obstacle avoidance for autonomous vehicles. Therefore, the performance of autonomous navigation systems improves [48].

5.2 Performance Comparison with Traditional Optimization Methods

The efficacy of the ABC algorithm in comparison to traditional optimization methods is crucial for determining its use. This comparison involves evaluating multiple aspects, such as the speed of convergence, the quality of the solution, the ability to handle fluctuations, and the computing efficiency. The ABC algorithm compared to the traditional optimization algorithm is given in Table 2.

ABC typically shows faster convergence in early iterations due to its exploitation and exploration balance. However, it may sometimes slow down in finding the exact global optimum compared to DE and PSO. ABC often provides high-quality solutions, especially for multimodal and high-dimensional problems, outperforming GA and SA. ABC demonstrates robustness across various test functions, maintaining consistent performance, unlike GA which might suffer from premature convergence [49]. ABC is computationally efficient, but DE can sometimes be more efficient due to its simpler operations.

Table 2 Comparison of genetic algorithms (GA) with PSO, SA, and DE

Parameters	GA	PSO	SA	DE
Convergence speed	Lower convergence speed	Lower convergence speed	Higher convergence speed	Higher convergence speed
Solution quality	Low-quality solutions	Low-quality solutions	Low-quality solutions	Low-quality solutions
Robustness	Not Robust	Not Robust	Not Robust	Not Robust
Computational efficiency	Less computationally efficient	Less computationally efficient	Less computationally efficient	More computationally efficient

5.3 Applications of ABC Algorithms

The ABC algorithm, which draws inspiration from the foraging activity of honeybees, has proven to be highly effective in many fields such as image recognition, natural language processing (NLP), and bioinformatics. The following are specific illustrations of the application of the ABC algorithm in certain domains:

5.3.1 Image Recognition

The ABC technique has been utilized in the field of image recognition for tasks including image segmentation, feature selection, and object detection. Following are certain applications of the ABC method in image recognition:

i. **Image Segmentation**: Image segmentation is an essential process in image recognition that entails dividing an image into separate regions or segments to facilitate its analysis [50]. Efficient segmentation facilitates the identification of objects and boundaries within an image. The ABC algorithm is used to search for threshold values that optimize a fitness function. This fitness function is usually based on characteristics such as inter-region contrast and intra-region homogeneity.
ii. **Feature Selection in Image Classification**: Feature selection is crucial for enhancing the performance of image classification algorithms by decreasing the data's complexity and choosing the most pertinent characteristics. The ABC method is employed to explore the optimal selection of features that maximize classification accuracy [51]. The utilization of the ABC-based feature selection strategy resulted in enhanced training efficiency and increased accuracy of image classification models, such as CNNs.
iii. **Object Recognition**: Object recognition is the process of accurately detecting and categorizing things present in an image. The ABC approach is applicable for optimizing the parameters of object recognition models [37]. The improved models demonstrated superior recognition accuracy and more rapid convergence

in comparison to models fine-tuned using conventional methods. It is possible to optimize the parameters of object recognition models, such as adjusting filter sizes and the number of layers in convolutional neural networks (CNNs).

5.3.2 Natural Language Processing (NLP)

The ABC approach optimizes model performance and feature selection in various NLP tasks. The following are specific applications of the ABC algorithm in NLP:

i. **Text Classification**: Text classification is the process of assigning text documents to predetermined categories, such as identifying spam, organizing topics, and analyzing sentiment. The utilization of the ABC-based feature selection method resulted in enhanced training efficiency and increased accuracy of classifiers, such as Support Vector Machines (SVMs) and Neural Networks [44].
ii. **Sentiment Analysis**: Sentiment analysis seeks to ascertain the sentiment conveyed in a text, whether it be good, negative, or neutral. The implementation of the ABC-based methodology resulted in enhanced precision and effectiveness of sentiment analysis models in comparison to conventional techniques [44].
iii. **Named Entity Recognition (NER)**: Named Entity recognition identifies and classifies named entities (people, organizations, locations). ABC-optimized NER models outperform baseline models in precision and recall [52].

5.3.3 Bioinformatics

The ABC algorithm has been effectively utilized in diverse bioinformatics applications to optimize computational procedures, enhance precision, and handle extensive biological datasets. These are specific applications of the ABC algorithm in the field of bioinformatics:

i. **Gene Selection in Microarray Data**: Gene selection entails the identification of a subset of pertinent genes using microarray data for the purpose of categorizing disorders, such as cancer [53]. The gene selection technique based on the ABC algorithm resulted in improved classification accuracy and decreased computational complexity in comparison to conventional methods.
ii. **Protein Structure Prediction**: Protein structure prediction seeks to ascertain the three-dimensional arrangement of a protein based on its sequence of amino acids. A refined ABC algorithm is employed to maximize the energy function utilized in protein structure prediction. The ABC approach enhanced the precision of forecasted protein structures and decreased the computational time needed [54].
iii. **Clustering of Biological Data**: Clustering is the process of categorizing related biological data points, such as genes or proteins, into clusters. The ABC-based clustering method yielded superior clusters with enhanced intra-cluster similarity and inter-cluster dissimilarity in comparison to conventional clustering techniques such as k-means and hierarchical clustering.

6 Hybridizing ABC with Other Metaheuristic Algorithms

In 1986, Fred Glover introduced the term "metaheuristic" [55] to describe a heuristic approach that does not include problem-specific characteristics. The effective searching mechanism of metaheuristic is accomplished by combining two search strategies: exploration (diversification) and exploitation (intensification) [56]. Metaheuristic algorithms are initially developed as a universal set of principles for getting problem-specific low-level heuristics, regardless of the challenge or domain. ABC swarm-based metaheuristic optimization strategies are inspired by the foraging behavior of honeybee colonies and are well-known for their ability to efficiently explore solution spaces [57]. Hybridizing bee colony optimization methods with other metaheuristic techniques is a prevalent approach to improve the search capabilities and efficiency of optimization algorithms for various real-life applications. Nevertheless, the fundamental ABC algorithm exhibits a sluggish convergence rate and has the propensity to become ensnared in local minima. Hybridization of this approach with other algorithms is necessary for these two reasons.

By integrating ABC with additional optimization approaches, one can attain an enhanced equilibrium between exploration and utilization of the search space. The method of hybridizing involves combining the principles, techniques, or components of two algorithms to generate a novel hybrid algorithm. Over 200 metaheuristic algorithms can be investigated and integrated with ABC [58]. Common strategies for integrating ABC with other optimization methods encompass the use of GA, PSO, simulated annealing (SA) algorithms [59], whale optimization algorithms (WO), and ACO algorithms [38]. The particular methodology employed is contingent upon the specific attributes of the problem at hand and the objectives of the optimization procedure.

6.1 Exploration of Combining ABC with Other Optimization Techniques

Hybridization of the ABC algorithms with other metaheuristics methods can be achieved in different ways which are as follows:

i. **At Initialization Level:** The ABC algorithms can be integrated with any other algorithm at the "*solution representation and initialization level*". The integration utilizes ABC algorithms to generate the initial population of solutions, while the accompanying algorithms progressively improve the result through iterations. For instance, when ABC is combined with SA algorithms, the ABC method is used to create an initial population of solutions, which are then enhanced iteratively utilizing the SA technique.
ii. **Exploration and Exploitation**: ABC algorithms can be integrated with other algorithms to facilitate both "*exploration and exploitation*". This strategy involves the utilization of both optimization algorithms to thoroughly explore

and exploit the search space. For example, When ABC is combined with PSO, the ABC algorithm uses bees for local exploitation, while the PSO algorithm uses particles to explore the search space globally.

iii. **Interactions among Operators of Algorithms:** The *"operator interactions"* of both algorithms add a new dimension to the functionality of hybridized ABC algorithms. The operators of different algorithms in the Operation Interaction technique are controlled by both ABC algorithms and combination algorithms. For instance, by integrating ABC with GA, the crossover and mutation operators from GA are utilized to generate novel solutions, while the ABC algorithm conducts local searches and assesses their fitness [60].

iv. **Termination and Convergence Criteria:** An ABC approach that has been hybridized can be configured to terminate at an optimal condition, resulting in improved convergence [61]. ABC algorithms cooperate with other metaheuristic algorithms to establish the halting condition of the algorithm. For example, integrating ABC with Tabu Search may need to incorporate tabu tenure and aspiration criteria to decide whether to terminate the search process.

Hybridizing ABC algorithms can be achieved not only through the methods mentioned but also can be done by including parametric optimization of both algorithms, utilizing meta-level information such as their strategies, and managing population selection for the next generation.

6.2 *Benefits and Challenges of Hybrid Approaches*

ABC algorithms, when combined with other optimization methods, demonstrate promise in developing resilient and flexible optimization algorithms capable of effectively addressing complex real-world problems. The utilization of the hybridized artificial colony of bees method in combination with other optimization strategies can result in significant benefits, but it also presents numerous challenges.

6.2.1 Benefits of Hybrid Approaches

Integrating the ABC method with additional optimization approaches or metaheuristic algorithms can provide numerous advantages which are as follows:

i. *Improved Exploration and Exploitation*: Integration of ABC with another algorithm, one can benefit from the advantages of both methods to efficiently navigate the solution space and target potential areas for improved solutions.

ii. *Improved Convergence:* Hybrid algorithms typically achieve faster convergence than standalone algorithms by integrating complementary search strategies to expedite the attainment of optimal or near-optimal solutions.

iii. *Robustness and Flexibility:* Hybridization enhances the optimization process by enabling it to be more resilient and flexible to various problem kinds and characteristics. The hybrid technique can assist mitigate the constraints or biases of algorithms, enhancing its versatility.
iv. *Search Diversity:* Hybridizing ABC algorithms can be achieved not only through the methods mentioned but also can be done by including parametric optimization of both algorithms, utilizing meta-level information such as their strategies, and managing population selection for the next generation.
v. *Quality of the Solution:* Incorporating ABC with other algorithms might enhance solution quality by using varied search tactics and exploration approaches that refine and enhance the overall quality of the produced solutions.

6.2.2 Challenges of Hybrid Approaches

Hybridizing the ABC algorithm with other methods presents numerous difficulties, representing a boundary where traditional optimization techniques meet contemporary computer approaches. The ABC algorithm is primarily based on utilizing the combined intelligence of artificial bees to effectively explore and exploit solution areas. Nevertheless, the incorporation of hybridization techniques presents complex obstacles which are as follows:

i. *Algorithmic Complexity:* Combining numerous algorithms into a hybrid strategy can augment the intricacy of the optimization process. These interactions, among diverse algorithms need to be managed and ensuring their compatibility might provide a challenge. The task of integrating many algorithms necessitates meticulous deliberation regarding their interactions and synergies [62]. A harmonious equilibrium between exploration and exploitation in the hybrid technique, without encountering conflicts between the constituent algorithms needs to be managed.
ii. *Implementation Overhead:* When implementing a hybrid algorithm, several algorithms and the parts that are combined with them must be integrated efficiently, which can add complexity and overhead to the implementation process.
iii. *Need Heavy Computational Resources:* Hybrid algorithms necessitate greater computer resources in comparison to solo alternatives due to the heightened intricacy and computational overhead linked with amalgamating various algorithms.
iv. *Lack of Transparency:* Hybrid algorithms may have less transparency than standalone techniques, which can make it challenging to comprehend and analyze their behavior.

6.3 *Case Studies Showcasing the Effectiveness of Hybrid Methods*

We have seen many studies employing hybridized ABC algorithms and other meta-heuristic algorithms, with promising outcomes. This section provides an overview of the hybrid approaches employed in different procedures. The hybridization approaches of significance are enumerated in Table 3 and elaborated upon in this section.

Even though there is a great deal of work to be done to develop a full metaheuristic framework, it is imperative that the essence of metaheuristic algorithms be kept, as well as the stochastic behavior that is inherent to them. The ABC method is a widely utilized swarm-based heuristic optimization algorithm that has been successfully applied to tackle numerous real-world situations.

Table 3 Case studies showcasing the effectiveness of hybrid methods

References	Algorithms	Approach	Limitations
Jatoth and Rajashekar [63]	ABC and GA	Proposed an innovative hybrid method for optimizing the proportional integral speed controller in a vector-controlled permanent magnet synchronous motor drive	Need heavy computational resources
Jia and Guo [64]	ABC and PSO	Proposed a hybrid approach and a hybrid methodology to efficiently address the resource-constrained project scheduling problem (RCPSP)	Need heavy computational resources
Shunmugapriya and Kanmani [38]	ABC and AC	Proposed a hybrid method that utilizes the exploitation of bees by ants to identify the optimal ant and the most suitable subset of features	Real-time applications that necessitate optimization are not utilized
Tang et al. [65]	ABC and WO	The ACWOA algorithm combines ABC techniques with chaotic mapping to enhance solution quality by preventing local optima	The threshold limit is determined through experimental means and does not possess any cognitive capabilities

7 Challenges and Future Research Directions

The hybridized ABC method is a highly effective optimization technique utilized for solving practical real-world problems. However, the integration of ABC encounters multiple challenges, and the use of developing approaches is crucial for achieving improved outcomes in the future. Although ABC possesses significant potential, it was evident to the researchers that certain enhancements are necessary to enhance its efficiency.

7.1 Addressing Challenges in Integrating ABC with Deep Learning

The integration of ABC algorithms with deep learning poses several obstacles due to the inherent disparities in their optimization methodologies and problem-solving tactics [66]. Some of the challenges of this integration are as follows:

i. **Dimensionality**: Deep learning models frequently possess parameter spaces with many dimensions, which render typical ABC techniques inefficient or ineffective due to the curse of dimensionality. ABC algorithms may encounter difficulties in successfully exploring and exploiting the extensive search space, resulting in slow convergence or subpar performance.
ii. **Complexity**: Deep learning models are characterized by their tremendous complexity, which is attributed to the presence of millions of parameters, nonlinear activation functions, and elaborate topologies [67]. ABC algorithms may have difficulties in efficiently traversing intricate terrains and discovering optimal solutions within practical time constraints.
iii. **Exploration–Exploitation Balance**: Attaining the optimal equilibrium between exploration, which involves diversification, and exploitation, which involves refinement, is essential for efficient optimization. ABC algorithms prioritize exploration, which can result in inferior solutions or premature convergence in deep learning problems that require exploitation to refine model parameters.
iv. **Rate of Convergence**: ABC algorithms generally exhibit slower convergence rates in comparison to gradient-based optimization approaches, particularly in high-dimensional domains [68]. Efficient optimization strategies are necessary for deep learning models to decrease training time and computational resources, as sluggish convergence poses a significant constraint.

7.2 Emerging Trends in Optimization for Deep Learning

The optimization techniques for deep learning are constantly changing to tackle the difficulties presented by intricate neural network frameworks, extensive datasets, and varied application fields [69]. There are several emerging optimization algorithms that can be coupled with deep learning. These methods include the following:

i. **Hybrid Optimization Techniques**: Hybrid optimization strategies leverage the advantages of various optimization methods to surpass their respective constraints. By integrating evolutionary algorithms like genetic algorithms or particle swarm optimization with gradient-based approaches, it is possible to increase the balance between exploration and exploitation and improve the resilience of optimization landscapes [70].
ii. **Adaptive Optimization Algorithms**: There is an increasing fascination with adaptive optimization algorithms that dynamically modify learning rates and other hyperparameters throughout training. Methods such as AdaGrad, RMSprop, Adam, and their variations have become popular due to their capacity to enhance the speed of convergence and the resilience of varied tasks and architectures [71].
iii. **Sparse and Low-Rank Optimization**: The utilization of techniques for sparse and low-rank optimization is becoming increasingly prominent, particularly for problems that involve large-scale deep learning models or environments with limited memory capacity [72]. Sparse regularization techniques, structured weight pruning, and low-rank factorization methods aid in decreasing the complexity of models, reducing memory usage, and lowering computational expenses while maintaining performance.
iv. **Stochastic Gradient Methods**: Stochastic gradient approaches, such as mini-batch gradient descent, as well as its variations like SGD with momentum and Nesterov momentum, continue to be essential in optimizing deep learning. Current developments prioritize the enhancement of convergence speed, stability, and generalization performance in situations including non-convex objectives or noisy gradients.
v. **Second-Order Optimization Methods**: Deep learning researchers are currently investigating the use of second-order optimization approaches, such as Newton's method and its variations [73]. These methods provide benefits in terms of accelerated convergence and improved handling of curvature information, but they also entail increased computing expenses. Current research is centered on creating effective approximations or combining several methods to make second-order techniques viable for deep learning tasks on a wide scale.
vi. **Robust Optimization and Adversarial Defense**: With the growing concern over adversarial attacks and model vulnerabilities, there's a rising interest in robust optimization techniques that enhance model robustness and resilience against perturbations. Adversarial training, robust optimization objectives, and regularization methods help improve the robustness of deep learning models against various forms of attacks.

7.3 Future Research Directions and Potential Advancements

Future research directions and potential advancements for ABC algorithms are likely to revolve around addressing key challenges, improving algorithmic efficiency, and expanding application domains. The research can be focused on the following areas:

i. **Enhanced Exploration and Exploitation**: ABC algorithms may prioritize enhancing the equilibrium between exploration and exploitation to enable more efficient exploration in intricate optimization landscapes. This may entail creating innovative methods for dynamically adapting exploration–exploitation tactics according to problem characteristics and evolutionary dynamics.
ii. **Parallel and Distributed Optimization**: Utilizing parallel and distributed computing architectures can greatly speed up the optimization process and provide scalability in dealing with large-scale optimization issues. Potential future possibilities could include the creation of parallel and distributed versions of ABC algorithms that utilize multi-core processors, GPU accelerators, or distributed computing frameworks to enable quicker convergence and enhanced scalability [74].
iii. **Multi-Objective and Constraint Optimization**: The extension of ABC algorithms to address constraints and multi-objective optimization issues continues to be a significant area of research [75]. Subsequent developments might center on the formulation of preference-based search strategies, Pareto-based ABC algorithms, and constraint-handling techniques—all of which would facilitate the efficient optimization of intricate, multi-criteria optimization tasks.

8 Conclusion

In conclusion, optimization is a critical process in various fields, aiming to find the best possible solutions to problems by adjusting decision variables to maximize or minimize an objective function within a set of constraints. Nature-inspired optimization algorithms offer powerful tools for solving complex optimization problems that traditional methods struggle with. The Artificial Bee Colony (ABC) algorithm, inspired by the foraging behavior of honey bees, plays a significant role in deep learning by aiding in feature selection, model structure optimization, hyperparameter tuning, and neural network training. ABC's ability to explore and exploit search spaces efficiently makes it an invaluable tool in enhancing deep learning models' performance. Hive learning, inspired by the collective intelligence of social insects, introduces a robust framework for distributed and collaborative learning. This approach is crucial for handling large-scale data and complex models, leveraging techniques like distributed learning, ensemble methods, and collaborative learning to improve model performance and efficiency.

Together, ABC algorithms and hive learning represent a convergence of nature-inspired optimization and advanced machine learning techniques, leading to more efficient, scalable, and robust AI systems. These concepts not only address current optimization challenges in deep learning but also open new avenues for future research and applications. As we continue to explore and integrate these techniques, the potential for breakthroughs in AI and machine learning grows, promising more sophisticated and capable systems.

The field continues to evolve, and several future research directions emerge, including (1) Combining ABC with other metaheuristic algorithms to leverage their complementary strengths, (2) Developing scalable versions of ABC for handling very large datasets and models, (3) Applying these techniques to a broader range of real-world problems to validate and refine their effectiveness, (4) Creating adaptive ABC algorithms that can dynamically adjust their parameters for different types of problems, and (5) Enhancing the theoretical understanding of ABC and hive learning to guide the development of more effective algorithms. By addressing these research directions, the integration of ABC algorithms and hive learning into deep learning optimization will continue to advance, driving innovations and enhancing the capabilities of AI systems.

References

1. Karaboga, D. (2005). An idea based on honeybee swarm for numerical optimization. Technical report-tr06, Erciyes university, engineering faculty, computer
2. Karaboga D, Akay B, Ozturk C (2007) Artificial bee colony (ABC) optimization algorithm for training feed-forward neural networks. In: Modeling decisions for artificial intelligence: 4th international conference, MDAI 2007, Kitakyushu, Japan, 16–18 Aug 2007. Proceedings, vol 4. Springer
3. Chong EK, Żak SH (2013) An introduction to optimization. John Wiley & Sons, vol 75
4. Gupta TK, Raza K (2019) Optimization of ANN architecture: a review on nature-inspired techniques. In: Machine learning in bio-signal analysis and diagnostic imaging, Elsevier, pp 159–182. https://doi.org/10.1016/B978-0-12-816086-2.00007-2
5. Raza K, Kohli M (2015) Ant colony optimization for inferring key gene interactions. In: Proceedings of 9th INDIACom-2015, 2nd international conference on computing for sustainable global development. IEEE, pp 1242–1246
6. Gupta TK, Raza K (2022) Optimization of Artificial Neural Network: a bat algorithm-based approach. In: Proceedings of 21st international conference intelligent systems design and applications, 13–15 Dec 2021. Lecture notes in networks and systems. Springer, vol 418, pp 286–295. https://doi.org/10.1007/978-3-030-96308-8_26
7. Arslan S, Ozturk C (2019) Multi hive artificial bee colony programming for high dimensional symbolic regression with feature selection. Appl Soft Comput 78:515–527
8. Rajasekhar A, Lynn N, Das S, Suganthan PN (2017) Computing with the collective intelligence of honey bees–a survey. Swarm Evol Comput 32:25–48
9. Karaboga D, Basturk B (2007) A powerful and efficient algorithm for numerical function optimization: artificial bee colony (ABC) algorithm. J Global Optim 39(3):459–471
10. Pham DT, Ghanbarzadeh A, Koc E, Otri S, Rahim S (2005) The bees algorithm: modelling foraging behavior to solve continuous optimization problems. In: Proceedings of the 2005 IEEE international conference on systems, man and cybernetics, vol 3, pp 2480–2487
11. Tang L, Zhang Y, Suganthan PN (2016) Benchmark functions for the CEC'2017 special session and competition on single objective real-parameter numerical optimization. Technical Report, Nanyang Technological University, Singapore

12. Yang XS, Deb S (2009) Cuckoo search via Lévy flights. In: World congress on nature & biologically inspired computing (NaBIC 2009). IEEE, pp 210–214
13. Karaboga D, Ozturk C (2011) A novel clustering approach: artificial bee colony (ABC) algorithm. Appl Soft Comput 11(1):652–657
14. Hao J, Sun X, Feng Q (2020) A novel ensemble approach for the forecasting of energy demand based on the artificial bee colony algorithm. Energies 13(3):550
15. Raza K, Singh NK (2021) A tour of unsupervised deep learning for medical image analysis current medical imaging. Bentham Sci 17(9):1059–1077. https://doi.org/10.2174/157340561 7666210127154257
16. Singh NK, Raza K (2023) TeethU2Net: a deep learning-based approach for tooth saliency detection in dental panoramic radiographs. In: Tanveer M et al (eds) 29th International conference on neural information processing (ICONIP 2022). CCIS 1794, vol 1794. Springer, pp 224–234. https://doi.org/10.1007/978-981-99-1648-1_19
17. Singh NK, Raza K (2022) Progress in deep learning-based dental and maxillofacial image analysis: a systematic review. Exp Syst Appl 199:116968. https://doi.org/10.1016/j.eswa.2022. 116968. (Elsevier)
18. Singh NK, Faisal M, Hasan S, Goshwani G, Raza K (2023) Dental treatment type detection in panoramic X-rays using deep learning. In: Proceedings of 22nd international conference intelligent systems design and applications (ISDA-2022), 12–14 Dec 2022. Lecture notes in networks and systems, vol 716. Springer, pp 25–33. https://doi.org/10.1007/978-3-031-355 01-1_3
19. Wani N, Raza K (2018) Multiple kernel learning approach for medical image analysis. In Dey N, Ashour A, Shi F, Balas E (eds) Soft computing based medical image analysis, Elsevier, pp 31–47. https://doi.org/10.1016/B978-0-12-813087-2.00002-6
20. Saravanan TM, Tamilarasi A (2016) Effective sentiment analysis for opinion mining using artificial bee colony optimization. Res J Appl Sci Eng Technol 12(8):828–840
21. Sheela SV, Radhika KR (2022) Iris recognition using multi objective artificial bee colony optimization algorithm with autoencoder classifier. Neural Process Lett 54(4):3489–3505
22. Baby Anitha E, Sivaprakash S, Velmurugan S, Saranya SS (2023) Hybrid artificial bee colony based neural network and dynamic threshold technique for predicting moving vehicle location and co-located objects. Sādhanā 48(2):68
23. Zhou B, Zhao Z (2023) An adaptive artificial bee colony algorithm enhanced by Deep Q-Learning for milk-run vehicle scheduling problem based on supply hub. Knowl-Based Syst 264:110367
24. Gao J, Lu Y, Qi J, Shen L (2019) A radar signal recognition system based on non-negative matrix factorization network and improved artificial bee colony algorithm. IEEE Access 7:117612–117626
25. Mosavi A, Ardabili, S, Várkonyi-Kóczy AR (2020) List of deep learning models. In: Engineering for sustainable future (pp. 202–214). Springer international publishing.
26. Raza K, Alam M (2016) Recurrent neural network based hybrid model for reconstructing gene regulatory network. Comput Biol Chem 64:322–334. https://doi.org/10.1016/j.compbiolchem. 2016.08.002. (Elsevier)
27. Glauner PO (2015) Comparison of training methods for deep neural networks. *arXiv preprint arXiv:1504.06825.*
28. Zaheer R, Shaziya H (2019) A study of the optimization algorithms in deep learning. Third international conference on inventive systems and control (ICISC), (pp. 536–539). Coimbatore, India.
29. Alamri NMH, Packianather M, Bigot S (2022) Deep learning: parameter optimization using proposed novel hybrid bees Bayesian convolutional neural network. Appl Artif Intell 36(1):2031815
30. Karaman A, Karaboga D, Pacal I, Akay B, Basturk A, Nalbantoglu U, Coskun S, Sahin O (2023) Hyper-parameter optimization of deep learning architectures using artificial bee colony (ABC) algorithm for high performance real-time automatic colorectal cancer (CRC) polyp detection. Appl Intell 53(12):15603–15620

31. Akay B, Karaboga D (2009) Parameter tuning for the artificial bee colony algorithm. In *Computational Collective Intelligence. Semantic Web, Social Networks and Multiagent Systems: First International Conference, ICCCI 2009, Wrocław, Poland, Proceedings 1* 5(7):608–619. October. Springer Berlin Heidelberg.
32. Loubiere P, Jourdan A, Siarry P, Chelouah R (2016) A sensitivity analysis method for driving the Artificial Bee Colony algorithm's search process. Appl Soft Comput 41:515–531. https://doi.org/10.1016/j.asoc.2015.12.044
33. Kang F, Li J, Xu Q (2009) Structural inverse analysis by hybrid simplex artificial bee colony algorithms. Comput Struct 87(13–14):861–870. https://doi.org/10.1016/j.compstruc.2009.03.001
34. Xiao S, Wang H, Wang W, Huang Z, Zhou X, Xu M (2021) Artificial bee colony algorithm based on adaptive neighborhood search and Gaussian perturbation. Appl Soft Comput 100:106955
35. Jadon SS, Tiwari R, Sharma H, Bansal JC (2017) Hybrid artificial bee colony algorithm with differential evolution. Appl Soft Comput 58:11–24
36. Arar ÖF, Ayan K (2015) Software defect prediction using cost-sensitive neural network. Appl Soft Comput 33:263–277. https://doi.org/10.1016/j.asoc.2015.04.045
37. Karaman A, Pacal I, Basturk A, Akay B, Nalbantoglu U, Coskun S, Sahin O, Karaboga D (2023) Robust real-time polyp detection system design based on YOLO algorithms by optimizing activation functions and hyper-parameters with artificial bee colony (ABC). Expert Syst Appl 221:119741. https://doi.org/10.1016/j.eswa.2023.119741
38. Shunmugapriya P, Kanmani S (2017) A hybrid algorithm using ant and bee colony optimization for feature selection and classification (AC-ABC Hybrid). Swarm Evol Comput 36:27–36. https://doi.org/10.1016/j.swevo.2017.04.002
39. Halim HZA, Moin NH (2022) Balance of exploration and exploitation in artificial bee colony for multi products inventory routing problem. Appl Math Sci 16(9):425–434. https://doi.org/10.12988/ams.2022.916842
40. Wang H, Wang W, Xiao S, Cui Z, Xu M, Zhou X (2020) Improving artificial bee colony algorithm using a new neighborhood selection mechanism. Inf Sci 527:227–240. https://doi.org/10.1016/j.ins.2020.03.064
41. Song X, Zhao M, Xing S (2020) A multi-strategy fusion artificial bee colony algorithm with small population. Expert Syst Appl 142:112921. https://doi.org/10.1016/j.eswa.2019.112921
42. Jiang Y, Qian H, Chu Y, Liu J, Jiang Z, Dong F, Jia L (2023) Convergence analysis of ABC algorithm based on difference model. Appl Soft Comput 146:110627. https://doi.org/10.1016/j.asoc.2023.110627
43. Kumar MA, Ramadevi Y (2022) A study: mammogram image segmentation and classification based on ABC algorithm and artificial neural networks. Adv Eng Serv (54/02)
44. Habeeb A, Otair MA, Abualigah L, Alsoud AR, Elminaam DS, Zitar RA, Ezugwu AE, Jia H (2022) Arabic text classification using modified artificial bee colony algorithm for sentiment analysis: the case of Jordanian dialect. In: Classification applications with deep learning and machine learning technologies. Springer International Publishing, Cham, pp 243–288. https://doi.org/10.1007/978-3-031-17576-3_12
45. Karaboga N (2009) A new design method based on artificial bee colony algorithm for digital IIR filters. J Franklin Inst 346(4):328–348
46. Gu T, Chen H, Chang L, Li L (2019) Intrusion detection system based on improved abc algorithm with tabu search. IEEJ Trans Electr Electron Eng 14(11):1652–1660. https://doi.org/10.1002/tee.22987
47. Hegazy O, Soliman OS, Salam MA (2014) LSSVM-ABC algorithm for stock price prediction. arXiv:1402.6366. https://doi.org/10.48550/arXiv.1402.6366
48. Lin TC, Chen CC, Lin CJ (2018) Wall-following and navigation control of mobile robot using reinforcement learning based on dynamic group artificial bee colony. J Intell Rob Syst 92(2):343–357. https://doi.org/10.1007/s10846-017-0743-y
49. Goudarzi S, Hassan WH, Anisi MH, Soleymani SA (2016) Comparison between hybridized algorithm of GA–SA and ABC, GA, DE and PSO for vertical-handover in heterogeneous wireless networks. Sādhanā 41:727–753. https://doi.org/10.1007/s12046-016-0509-4

50. Cuevas E, Sención-Echauri F, Zaldivar D, Pérez M (2013) Image segmentation using artificial Bee colony optimization. In: Handbook of optimization: from classical to modern approach, pp 965–990. https://doi.org/10.1007/978-3-642-30504-7_38
51. Agrawal V, Chandra S (2015) Feature selection using Artificial Bee Colony algorithm for medical image classification. In: 2015 eighth international conference on contemporary computing (IC3). IEEE, pp 171–176. https://doi.org/10.1109/IC3.2015.7346674. (Aug 2015)
52. Gharagozlou H, Mohammadzadeh J, Bastanfard A, Ghidary SS (2022) RLAS-BIABC: a reinforcement learning-based answer selection using the bert model boosted by an improved ABC algorithm. Comput Intell Neurosci. https://doi.org/10.1155/2022/7839840
53. Alshamlan H, Badr G, Alohali Y (2015) mRMR-ABC: a hybrid gene selection algorithm for cancer classification using microarray gene expression profiling. Biomed Res Int. https://doi.org/10.1155/2015/604910
54. Tordai H, Suhajda E, Sillitoe I, Nair S, Varadi M, Hegedus T (2022) Comprehensive collection and prediction of ABC transmembrane protein structures in the AI era of structural biology. Int J Mol Sci 23(16):8877. https://doi.org/10.3390/ijms23168877
55. Glover F (1986) Future paths for integer programming and links to artificial intelligence. Comput Oper Res 13(5):533–549. https://doi.org/10.1016/0305-0548(86)90048-1
56. Blum C, Roli A (2003) Metaheuristics in combinatorial optimization: overview and conceptual comparison. ACM Comput Surv (CSUR) 35(3):268–308. https://doi.org/10.1145/937503.937505
57. Kaya E, Gorkemli B, Akay B, Karaboga D (2022) A review on the studies employing artificial bee colony algorithm to solve combinatorial optimization problems. Eng Appl Artif Intell 115:105311. https://doi.org/10.1016/j.engappai.2022.105311
58. Sarumaha YA, Firdaus DR, Moridu I (2023) The application of artificial bee colony algorithm to optimizing vehicle routes problem. J Inf Syst Technol Eng 1(1):11–15. https://doi.org/10.61487/jiste.v1i1.9
59. Chandrashekhar M, Dhal PK (2024) Multi-objective economic and emission dispatch problems using hybrid honeybee simulated annealing. Meas Sens 101065
60. Mumtaz J, Minhas KA, Rauf M, Yue L, Chen Y (2024) Solving line balancing and AGV scheduling problems for intelligent decisions using a Genetic-Artificial bee colony algorithm. Comput Ind Eng 189:109976. https://doi.org/10.1016/j.cie.2024.109976
61. Ye T, Wang H, Zeng T, Omran MG, Wang F, Cui Z, Zhao J (2024) An improved two-archive artificial bee colony algorithm for many-objective optimization. Exp Syst Appl 236:121281.https://doi.org/10.1016/j.eswa.2023.121281
62. Durgut R (2021) Improved binary artificial bee colony algorithm. Front Inf Technol Electron Eng 22(8):1080–1091. https://doi.org/10.1631/FITEE.2000239
63. Jatoth RK, Rajasekhar A (2010) Speed control of PMSM by hybrid genetic artificial bee colony algorithm. In *2010 International Conference on Communication Control and Computing Technologies* (pp. 241–246). IEEE. October.
64. Jia Q, Guo Y (2016) Hybridization of ABC and PSO algorithms for improved solutions of RCPSP. *Journal of the Chinese Institute of Engineers*, *39*(6):727–734.
65. Tang C, Sun W, Xue M, Zhang X, Tang H, Wu W (2022) A hybrid whale optimization algorithm with artificial bee colony. Soft Comput 26(5):2075–2097. https://doi.org/10.1007/s00500-021-06623-2
66. Akay B, Karaboga D, Gorkemli B, Kaya E (2021) A survey on the artificial bee colony algorithm variants for binary, integer and mixed integer programming problems. Appl Soft Comput 106:107351. https://doi.org/10.1016/j.asoc.2021.107351
67. Babalola AE, Ojokoh BA, Odili JB (2020). A review of population-based optimization algorithms. In: 2020 International conference in mathematics, computer engineering and computer science (ICMCECS). IEEE, pp 1–7. (Mar 2020)
68. Ezazi F, Mallah MH, Sabet JK, Norouzi A (2020) A new method for multicomponent mixture separation cascade optimization using artificial bee colony algorithm. Prog Nucl Energy 124:103371

69. Kumar A, Kumar D, Jarial SK (2017) A review on artificial bee colony algorithms and their applications to data clustering. Cybern Inf Technol 17(3):3–28. https://doi.org/10.1515/cait-2017-0027
70. Kiliçarslan S, Dönmez E (2023) Improved multi-layer hybrid adaptive particle swarm optimization based artificial bee colony for optimizing feature selection and classification of microarray data. Multimed Tools Appl 1–23. https://doi.org/10.1007/s11042-023-17234-4
71. Hancer E (2020) Artificial bee colony: theory, literature review, and application in image segmentation. In: Recent advances on memetic algorithms and its applications in image processing, pp 47–67. https://doi.org/10.1007/978-981-15-1362-6_3
72. Ouhame S, Hadi Y, Arifullah A (2020) A hybrid grey wolf optimizer and artificial bee colony algorithm used for improvement in resource allocation system for cloud technology
73. Chen L, Wu T, Wang Z, Lin X, Cai Y (2023) A novel hybrid BPNN model based on adaptive evolutionary Artificial Bee Colony Algorithm for water quality index prediction. Ecol Ind 146:109882. https://doi.org/10.1016/j.ecolind.2023.109882
74. Bai Y, Zhang C, Bai W (2023) A two-level parallel decomposition-based artificial bee colony method for dynamic multi-objective optimization problems. Appl Soft Comput 147:110741
75. Zheng QQ, Zhang Y, He LJ, Tian HW (2023) Discrete multi-objective artificial bee colony algorithm for green co-scheduling problem of ship lift and ship lock. Adv Eng Inform 55:101897. https://doi.org/10.1016/j.aei.2023.101897
76. Jia Q, Guo Y (2016) Hybridization of ABC and PSO algorithms for improved solutions of RCPSP. J Chin Inst Eng 39(6):727–734. https://doi.org/10.1080/02533839.2016.1176866
77. Pan Q-K et al (2011) A discrete artificial bee colony algorithm for the lot-streaming flow shop scheduling problem. Inf Sci 181(12):2455–2468

An IoMT Enabled Iterative Artificial Bee Colony Approach Using Federated Learning for Detection of Heart Disease

Nayyar Ahmed Khan, Md. Mobin Akhtar, Ahmed Masih Uddin Siddiqi, Khan Asif Rashid, Sivaram Rajeyyagari, Mohammad Nadeem Khalid, and Mohammad Ahmad

1 Introduction

The introduction of a new era in medical diagnosis started with using IoT devices in diagnosis and therapy. Several new parameters and symptoms are rising with the growth of IoT devices. These devices are small and usually consume little power. It makes the work simpler and thus provides a good quality of life and diagnosis. The biomedical data for individual patients to a group of people can be managed with the help of these small devices. The connectivity of the devices with the network enables the flow of information from one node to another without any human intervention. Massive amounts of biomedical data can flow from one point to another with straightforward steps [3]. The use of modern technology like wearable devices, smart watches, sports bands, etc., enables the detection of problems associated with

N. A. Khan (✉) · K. A. Rashid · S. Rajeyyagari
Department of Computer Science, College of Computing and Information Technology, Shaqra University, Shaqra, Saudi Arabia
e-mail: nayyar@su.edu.sa

Md. M. Akhtar
Department of Basic Science, College of Applied Medical Sciences, Riyadh Elm University, Riyadh, Saudi Arabia

A. M. U. Siddiqi
Department of Computer Engineering and Application, Mangalayatan University, Aligarh, India

M. N. Khalid
Department of Electro-Mechanical Engineering Technology, Institute of Applied Technology, Abu Dhabi Polytechnic University, Abu Dhabi, United Arab Emirates

M. Ahmad
Department of Computer Science, College of Science and Humanities-Dawadmi, Shaqra University, Al Duwadimi, Saudi Arabia

K. Raza (ed.), *Solving with Bees*, Springer Tracts in Nature-Inspired Computing,
https://doi.org/10.1007/978-981-97-7344-2_6

the health and symptoms of the physical body. Healthcare refers to the early diagnosis of the symptoms, thereby providing proper medication or support to retrain from the ailment [12]. With the advancement of technology, it is possible to transmit bulk data from one point to another to provide an efficient method of sophisticated healthcare.

As per the World Health Organization (WHO), it is estimated that approximately 18.5 million people die because of cardiovascular diseases every year [16]. Various conditions inside the human body affect the functioning of the heart and may lead to diseases connected to cardiovascular abnormalities. There are multiple factors that are responsible for these abnormalities related to genetics, use of products like caffeine, overconsumption of alcohol, stressed life, lack of fitness, variation in blood pressure, obesity due to improper living trends, pre-existing issues related to heart and cholesterol, etc. [19]. Patients struggling with cardiovascular problems can be saved if they are detected at early stages. Due to the lack of identification of the disease caused by heart malfunction, the fatality rate rises, and the probability of tracking the disease. Machine learning explores various factors and methods by which the diagnosis of heart-related issues can be made [14]. Classifying multiple factors and features related to the diagnosis can lead to adequately protecting the ailment. It is thus possible to identify and protect the problems associated with the heart with the help of machine learning techniques. Several researchers have used approaches to suggest these problems with the help of random forest or linear regression models in machine learning. Li et al. [22] reported a system capable of identifying these problems based on a classification algorithm.

The machine learning models are trained with the help of the data collected from the healthcare units related to a patient's health status. In the present scenario, obtaining the data related to the patient's ailment in real time is tough. Individual data secrecy and privacy issues are the most critical factors that lead to problems associated with machine learning identification and classification. Despite the enormous amount of data in various organizations, sharing the information remains a barrier. Several laws by the European Union [32], Chinese security, consumer privacy in the United States of America and California [13], etc., made it challenging to track the data for further studies. Due to a lack of data from real-time objects, a new computational paradigm called federated learning was recently proposed by Google [27]. The main idea of this new computational shift is a collaborative mechanism in which information from various nodes flows through a network and provides a comprehensive set of data that can be analyzed using machine learning. As suggested by Alshahrani et al. [6], we use Federated learning to apply machine learning to the bulk datasets for optimization to reduce privacy issues and diagnose heart-based diseases. This research primarily focuses on developing an architecture that can provide better features using machine learning algorithms.

The algorithm we selected for this study is the Iterative Artificial Bee Colony (I-ABC). It is a user-friendly and straightforward algorithm to implement and integrate various parameters. We have compared the results from our proposed architecture with existing complex algorithms such as particle swarm optimization and genetics algorithms. As suggested by Alshahrani [7], particle swarm optimization can be used to identify suitable candidates from the dataset. However, artificial bee colony (ABC) optimization [18] refers to a more robust and eligible technique to identify the possible results. The system designed in this framework uses machine learning to identify the prediction of heart diseases in the medical and healthcare unit. The contribution of our paper refers to the following:

- The I-ABC is a technique proposed in the architecture in which the information flows from various client nodes toward a centralized computational unit responsible for disease prediction using Federated Learning (FL).
- We developed the Federated learning integration with the machine learning algorithm of the ABC optimization technique.
- To validate the complete system, we use some third-party datasets and figure out the dependency and accuracy of the system.
- Before then, try to identify the comparison between the existing systems and the proposed architecture given in this study.
- Finally, the focus is to identify the prediction capability of the system to protect real-world heart disease issues. Classification, optimization, errors, and efficiency are the prime factors investigated in this study.

The generic approach for various researchers across the world focuses mainly on the collection of data and analysis. The data concerned with a specific illness is collected from multiple healthcare units. These healthcare units are responsible for storing and preserving a person's data. Privacy and the security of data are the primary concerns in the field of data analytics. It is tough to obtain data for the patient from the medical healthcare units with their situation. Most human beings are reluctant to provide the data for analysis due to privacy issues. The main problem arises with the need for more data due to the rules associated with individual privacy by various countries and organizations. To overcome the problem of data security and privacy, we integrate the concept of Federated learning in this architecture. The technique involves collecting data from various individual nodes that flow from IoT devices and reach the centralized server for execution. We optimize the data regarding collection, dressing, management, analysis, and result publication.

2 Literature Review

Several results are discussed by Almalki et al. [4] for the Federated learning environment, they depend on the hardware, software, technology, protocol, and various platforms concerned. Various advantages and disadvantages, including the challenges and the issues in Federated learning, are presented by AbdulRahman et al. [2]. The authors discussed several edge and fog computing techniques in a Federated learning environment [29]. The use of individual patient's electronic health records to discuss and protect the issues related to the diseases was presented in [10]. However, there has been considerable growth in the electronic health records sector and the distribution of these records by various agents in the real-time environment. The authors

[29] discussed several algorithms that use support vector machines and a Federated learning environment. The proposed scenario produced more accurate results than the previous techniques. Yet another exciting work by Nguyen et al. [28] revealed an algorithm that could make intelligent selections at the individual device levels in the model. The approximation and the communication for the best possible optimal solution were protected at the device level to increase the convergence rate. It enhances the model's accuracy but uplifts the hardware required in the study at the individual node level. In the survey by Ma et al. [23], an algorithm for predicting diseases based on the weights attached to the individual vectors was proposed. A couple more researchers have also used federated learning to indicate and identify the COVID-19 virus.

The authors used traditional machine learning algorithms in Abdul Salam et al. [1] where they could more accurately protect the disease associated with the data prescribed in the machine learning algorithm. The authors used Several datasets, including chest X-rays and patient reports from radiographic images. The dataset consists of several more factors, such as pneumonia value and standard images, to train the model to a higher level. In Cheng et al. [11], federated learning was used with the blockchain to provide security, and privacy is demonstrated. People accept blockchain verifications and hash function encryption in terms of secrecy and confidentiality of the data collected for an individual patient. The idea proposed by Almalki et al. [4] used blockchain to preserve the confidentiality of information and safety from data leakage across IoT devices. It ensures that the data flowing from the healthcare unit will not suffer a security breach and will reach the exact location whenever required. Secrecy has become an integral part of the blockchain system that ensures the use of digital contracts for proper information safety. Data from several empty I/O devices is collected with the help of cloud-based services. Almost all cloud service providers provide information collection with the help of the API.

Data from the IoT devices flows from the existing node to the cloud environment. This data is saved across the servers from the sensors so that predictions can be made based on the data analytics. However, privacy is a significant concern, as per [15]. In this scenario, an ABC optimization technique is used to manage the information for various IoT-based data applications [9]. This technique handled the workload for the data in the IoT application. Yet another model proposed by Karaboga [17] comprises the optimization for the computational environment in which data protection inside the edge computing device was done. These edge-oriented devices could execute ABC optimization at the node level to process the data quickly at the server. The information that flows from individual nodes was already dressed and cleaned for final execution. The long short-term memory (LSTM) technique has been applied in several studies for the edge computing scenario proposed by Zaman et al. [35]. In this technique, the prediction was done by offloading the IoT data from the individual node to the server. The surveys offered by several researchers [25, 31, 34] provide an extensive way to identify open datasets that can be used to categorize and identify COVID-19 from X-ray images and the patient's CT scans. This data was diverse and comprised several features related to the reporting and diagnosis.

Several more methods were also implemented by various researchers for the heart disease prediction environment that comprises machine learning algorithms like ant lion optimization [20], Decision trees [34], Swarm optimization, and objective functions for the new Row-Fuzzy system [26]. Several more researchers have implemented architecture related to monitoring and prediction with the help of neural networks for IoT-enabled devices [20, 21, 30]. Almost all the methods can improve the accuracy, performance, and classification of pre-existing methods for prediction. The use of genetic algorithms and artificial intelligence for several electrocardiogram ECG analyses was also implemented [8]. Almost all the strategies discussed in this section relate to heart disease reduction and data optimization. Every technique has some potential, along with several challenges as well. The idea proposed in this work comprises a model designed for integrating Federated learning with the I/O empty devices and ABC optimization to predict the accurate results of heart disease prediction. Section 3 explains the system's complete methodology and the material required for the development.

3 Materials and Methods

IoMT devices make it possible to collect real-time data from various healthcare unit patients. Based on the data collected from these devices, a proper diagnosis can be made with the help of machine learning algorithms. Due to privacy laws in various countries, it is tough to capture information from individual patients across different healthcare units around the globe. To enable the learning and working paradigms, we implement Federated learning to achieve this goal. The sensor devices connected across the patient's body share the data from various nodes through a secure network. In the healthcare unit, security and privacy can be overcome with the help of Federated learning to predict heart diseases. We apply Federated learning in Association with a machine learning algorithm to provide a secure data collection and analysis mechanism. The ABC algorithm optimization technique makes identifying the most suitable candidate in the complete dataset possible.

As per Fig. 1, the ABC algorithm is one of the most exciting optimization algorithms in machine learning. It is simulated with the working of bees in a high. The Queen bee is selected based on the performance of the scout bees, onlooker bees, and the finally employed bee. The parameters from the dataset are the bees, in this case, in the simulation. The population is checked to find out the most appropriate value. Once the possible solution from the dataset is identified, it is selected as the Queen bee. However, if the selected feature is unrelated to the Queen bee, the iteration retakes place relative to finding the primary bee value. Once the final solution is received, the best possible result is restored as the most eligible candidate among all the features available. This algorithm is more relevant to optimizing the datasets by finding the most eligible feature.

We deploy a dataset with proper training and testing to ensure the model works appropriately and tests under different inputs. The method adopted in this study uses

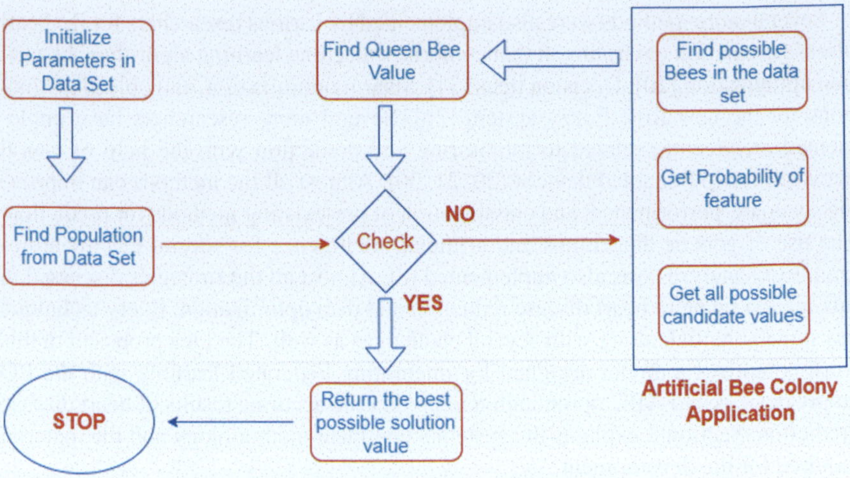

Fig. 1 Artificial bee colony algorithm workflow in the proposed study

machine learning as the backbone. As depicted in Fig. 2, the dataset is selected for the testing and training of the model. The preprocessing of the data takes place on the server side. Once the data is pre-processed in a proper format, the features related to identifying heart diseases are extracted from the dataset. The selected features are loaded into the ABC optimization technique deployed in machine learning for datasets. Once the specified features are packed into the optimization technique, they are processed.

The feature selection occurs in the previous stage before the optimization—machine learning emphasizes identifying the features and processing the data based

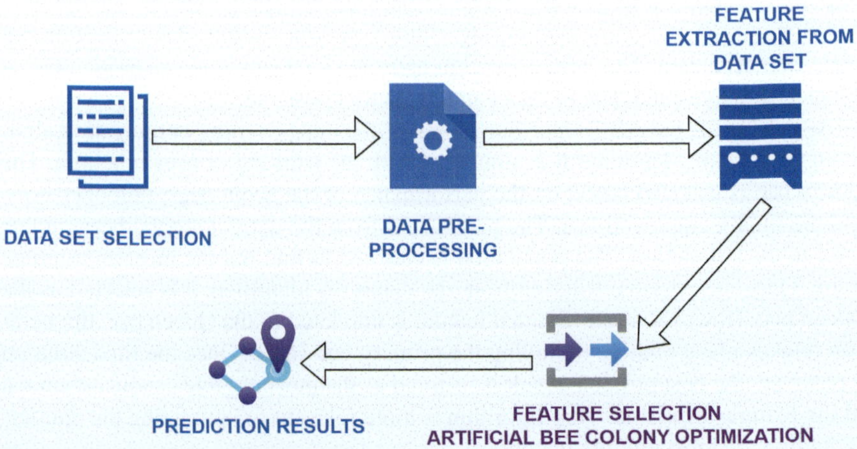

Fig. 2 Core Functionality of the model proposed using machine learning

on specific rules. The ABC optimization technique is beneficial for predicting the results depending on several features selected in this algorithm.

4 Dataset Description

We have trained and tested the model for ABC optimization using the UCI dataset (University of California and Irvine). This data set contains around 302 records and 76 attributes. The values of the risk associated with different parameters in this dataset are represented in the table. The values are appropriately coded with Boolean values in the dataset to predict the solution. We applied the dataset to our ABC optimization algorithm to identify the results. The prediction is based on several factors per the encoded values, including alcohol consumption, diabetes detection, electrocardiography technique values, fasting blood sugar, heart can, smoking habits, diet parameters, cholesterol value, heartbeat, and diseases related to the heart in the past tense.

5 IoMT Clients I-ABC Technique

The algorithm prescribed in this study comprises an iterative ABC optimization technique. As per the ABC optimization, the central ideology returns to identifying the scout bees, employed bees, and all the onlooker bees. The identification of the leading candidate bee takes place using this technique. The model proposed in this study implements the iterative method for identifying the candidate bee depending upon the Federated learning rules. Initially, the data from the devices flows from the network towards the ABC algorithm for the prediction.

As shown in Fig. 3 above, the predicted value is stored inside the rules of the trained model. The Federated learning rule is updated based on the results identified from the first iteration of the ABC algorithm. The model is further qualified based on the rules from the individual nodes. The candidate value is determined because the ABC algorithm is updated to aggregate the federated learning rules and pass them to the trained model. The model further integrates the rule to the individual nodes. During the federated learning phase, the individual nodes get a replica of the rules prescribed in this study. These integrated rules are further passed on to the ABC algorithm in the next iteration. Every time the algorithm executes itself, the rules get updated and provide a more sophisticated output depending upon the input given by the regulations. The main aim of this strategy is to identify the most employed value as a result. The fitness function responsible for determining the optimal selection during the classification makes all the IoMT client sites eligible for identification of the most employed bee. The algorithm shown below generalizes the working of the optimization technique:

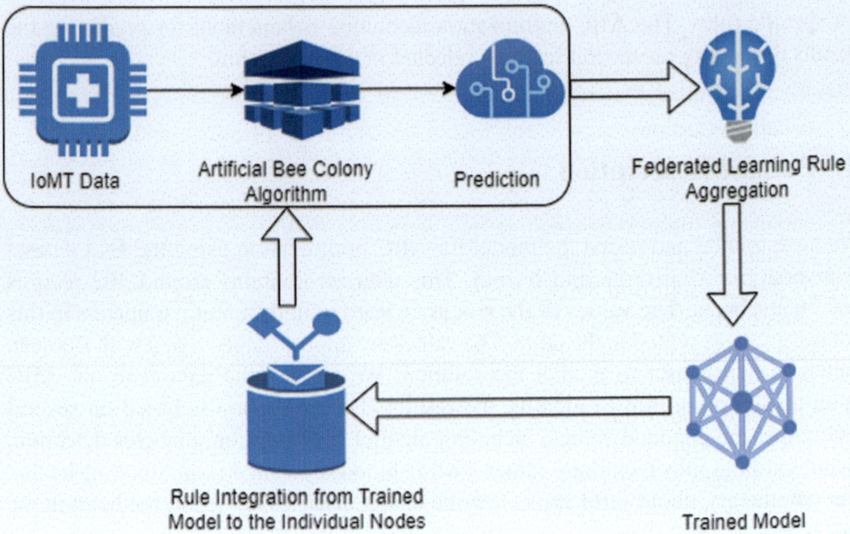

Fig. 3 Iterative artificial bee colony algorithm in the proposed study (inspired by [5])

The algorithm prescribed in this study contains essential equations referred from [26]. The first equation identifies the client node from the pool of several nodes connected to the analysis Centre. This equation allows us to select a random value of the client node, which will be analyzed to identify the most appropriate feature.

Algorithm I: Optimization from the Iterative ABC Algorithm

```
Input: IoMT Data Read takes place
I-ABC_Processing (node)
  For: each node value Repeat:
        Step I: Search the most viable bee value
        Step II: Calculate the onlooker bees
        Step III: Finalize the employed bee
        Step IV: Save the result of the feature.
  End For
For: each node update:
        I-ABC_Processing (new Node value)
        If (End of Data Set)
                End
        Else
                Continue
End For
```

$$X_{ni} = \text{li} + \frac{[(\text{rand}(-1, 1) + 2i - 1) * (\text{Ui} - \text{li})]}{2\text{NP}}$$

where X_{ni} refers to the values of the IoMT Healthcare sites, Ui and Li represent the upper and lower bounded values for the concerned values in the IoMT data received for a feature.

The selection of the candidate from the IoMT values is done with the help of equation number two [26]. This equation uses a fitness function, which is added to the randomly selected value of the candidate nodes.

$$C_{ni} = \begin{cases} F_{ni} + \text{rand}(X_{ni} - X_{pi});\ \text{if} i = i + 1 \\ X_{pi};\ \text{if} i \neq i + 1 \end{cases}$$

where C_{ni} is the most eligible candidate value of the bee selected from the data set of the IoMT Data received. The function F_{ni} refers to the fitness function value which is added to the random value. The values of the candidate values are X_{ni} and X_{pi}

After selecting several candidates from the node, the most appropriate candidate is identified and connected depending on the value of the fitness function. Equation three below represents identifying the most suitable candidate for the fitness function given by Mateen Yaqoob et al. [26].

$$C_{nio} = \frac{F_{ni}(Xn)}{\sum_{i=1}^{n}(Fm)(Xn)}$$

where C_{nio} is the most appropriate candidate feature/bee value employed as queen bee. The value of F_{ni} is the fitness function defined as given below:

$$F(n) = \begin{cases} \dfrac{1}{1 + Fi}, & \text{if } Fi \geq 0 \\ 1 + \text{abs}(Fi), & \text{if } Fi < 0 \end{cases}$$

The fitness function represented above ensures the eligible candidate's calculation based on the candidate's available fitness value [26]. The system presented in this study refers to calculating all four equations stated above to identify the most eligible candidate from the existing features in the dataset values.

6 Design of the Proposed Model

The proposed model presents an environment for collecting data from various nodes or healthcare units inside the system. The data is fetched in a Federated learning environment to ensure secrecy and privacy as shown in Fig. 4. The proposed framework collects data related to heart diseases inside a healthcare vicinity. The data from all the sites is covered using federated learning, and it is used to perform the feature selection along with the optimization of the data from the trained model. The data remains at the node where it is generated, but it is only transferred with more accuracy, and the classification occurs with the most minor errors.

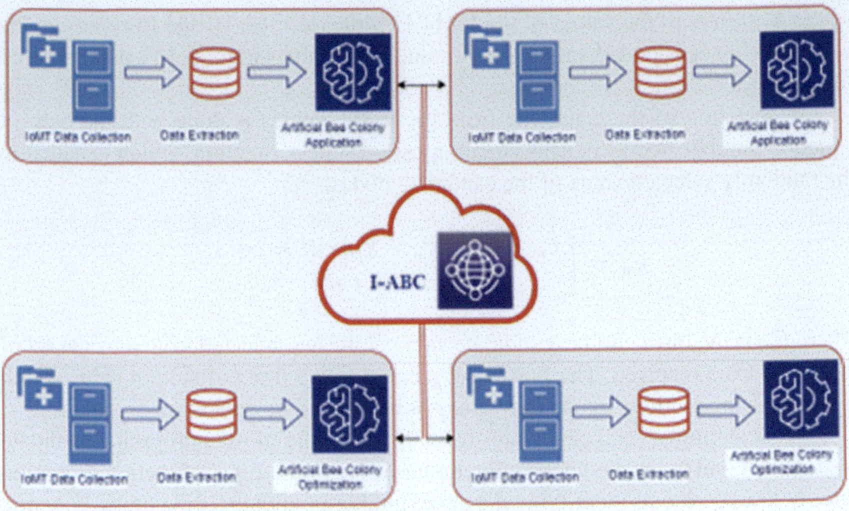

Fig. 4 Federated learning environment for the proposed model FL-ABCO

The proposed architecture intends to use the ABC algorithm in the healthcare unit. The most suitable candidate is passed across the secure network. The central cloud covers all the information and stores the data to identify the most appropriate value. The site where all the data is collected uses the algorithm, I proposed above to determine the most eligible candidate value. The queen bee is selected with the help of a Federated learning environment to ensure the complete working of the proposed system.

7 Experimental Setup

The performance evaluation for the proposed architecture is done with the help of the data set available from the UCI open-source data. We conducted the simulation based on the proposed architecture for the I-ABC optimization techniques registered in this study. The simulation took place for nearly 1500 execution rounds using the Py-Torch libraries of the Python Programming Language for machine lsearning. The competition was on a standard machine with an Intel Core i-5 processor. All experiments that are conducted in the execution make use of the default values of the data sets. In this study, we have considered three client nodes sharing the data from individual sensor devices as per the data set.

8 Results and Discussion

We have collectively tested the framework and the data set to measure the system's efficiency proposed in this study. The training and testing are done to analyze the accuracy of the standard Federated learning methodologies. We have implemented two state-of-the-art methods related to Federated learning. Information from the client notes is expected to flow as part of the two matters for the Federated learning environment.

The prediction for heart disease was the core idea behind the development of this model. The focus was to enhance the production capability of the system and diagnose heart diseases based on the machine learning algorithm. We have considered two state-of-the-art methods in a Federated learning environment, FEDSGD and FEDMA, for comparison with our model. The framework proposed in this study achieved an accuracy level of 91.1% in the total execution of the framework. Since we offer the I-ABC optimization technique applied to the data set, the accuracy is higher than that of traditional methods. The methodology used in the state-of-the-art techniques comprises client-based models, which leads to more convergence from the actual cloud model. In our case, we have applied an iterative approach to improve the learning rate for the model compared to the two traditional methods.

The model's accuracy is compared with the state-of-the-art methods FEDSGD [24] and FEDMA [33] as shown in Fig. 5. The system's processing takes more than 1500 rounds to attain substantial accuracy. The system's performance proposed in this study is based on calculating all the parameters required for the complete data set features.

We calculated performance metrics such as accuracy, sensitivity, precision, and specificity as shown in Table 1, reflecting the superiority of the proposed model over other state-of-the-art methods in Federated learning. The proposed architecture comprises several highlighting features, including enhanced accuracy and our iterative methodology to improve the data set and its values. The complete performance analysis of the prescribed method gives the proper prediction of heart diseases from the data set. Precision is more in predicting the supplied data set to the algorithm presented in this work. These error specificity and sensitivity values will be beneficial for a better selection of features in the existing model. The classification errors are also minimal in the proposed model, and the efficiency of the model is higher.

9 Conclusion

In this study, we have proposed a secure and robust model for the Federated learning framework in terms of the prediction of heart diseases. This framework suggested in this study uses a new iterative artificial bee colony (I-ABC) algorithm technique to predict heart-based diseases. The main aim of the proposed chapter in this study is to identify a mechanism responsible for proper, accurate heart disease prediction.

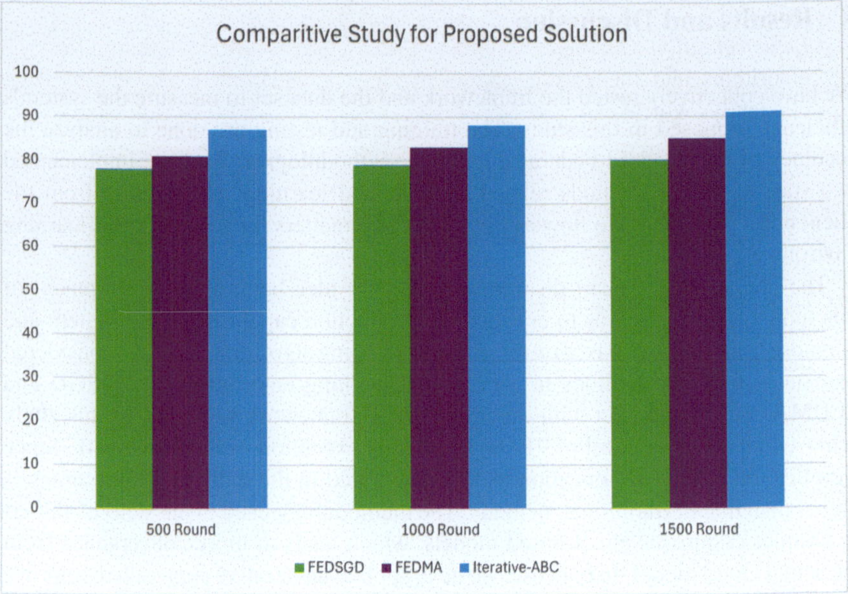

Fig. 5 Comparison of the proposed Iterative ABC method with other state-of-the-art methods

Table 1 Comparative values for all the parameters

Method	Accuracy (%)	Precision (%)	Specificity (%)	Sensitivity (%)
I-ABC (Proposed method)	91.1	90.2	52.7	89.7
FEDSGD [24]	90.5	89.89	53.1	89.9
FEDMA [33]	90.1	90.02	52.3	90.5

The study focuses on creating our model, which uses several sensor bay devices to record information from a healthcare unit and stream it after applying an I-ABC optimization technique. Federated learning is used in this regard, together with the information from various network nodes, to preserve the secrecy and privacy of individual data. Several factors are studied in this study to ensure that the proposed architecture has more accuracy and precision. We investigated the article concerning the existing state-of-the-art methods in a Federated learning environment. The core idea of prediction acquires 91.1% accuracy with 90.2% precision. The consolidated error that might occur during the computation is as low as 18.01%. Several other factors, like sensitivity, specificity, and F value, were calculated and found to be considerably low compared to traditional methods. Our prescribed model has several challenges and limitations as well. We have tested the model with three client nodes only, and it is expected that the model still needs to be updated in terms of many client sites. We hope to enhance the testing and training of the model to more client sites soon. The diagnosis of heart-related diseases is indeed one of the most important problems associated with healthcare across the world. Enhancing and modifying the model proposed in this study will help humanity produce better and more efficient results. Several more diseases, like breast cancer, diabetes, lung infection, etc., can also be treated with the help of a similar approach in the future. The limitations are restricted due to the unavailability of computational resources for a greater number of clients and federated locations from healthcare. The dataset used is an open dataset which needs to be replaced with a real-time dataset to provide more challenging results. We aim to focus on increasing the number of client locations and real-time clinical data from the healthcare vicinity to make the system more robust.

References

1. Abdul Salam M, Taha S, Ramadan M (2021) COVID-19 detection using federated machine learning. PLoS ONE 16(6):e0252573
2. AbdulRahman S, Tout H, Ould-Slimane H, Mourad A, Talhi C, Guizani M (2020) A survey on federated learning: the journey from centralized to distributed on-site learning and beyond. IEEE Internet Things J 8(7):5476–5497
3. Al-Turjman F, Nawaz MH, Ulusar UD (2020) Intelligence in the internet of medical things era: a systematic review of current and future trends. Comput Commun 150:644–660
4. Almalki J, Al Shehri W, Mehmood R, Alsaif K, Alshahrani SM, Jannah N, Khan NA (2022) Enabling blockchain with IoMT devices for healthcare. Information 13(10):448
5. Almalki J, Alshahrani SM, Khan NA (2024) A comprehensive secure system enabling healthcare 5.0 using federated learning, intrusion detection and blockchain. PeerJ Comput Sci 10:e1778
6. Alshahrani SM, Almalki J, Alshehri W, Mehmood R, Albahar M, Jannah N, Khan NA (2023) Systematic survey on big data analytics and artificial intelligence for COVID-19 containment. Comput Syst Sci Eng 47(2)
7. Alshahrani SM, Khan NA, Almalki J, Al Shehri W (2022) URL phishing detection using particle swarm optimization and data mining. Comput Mater Contin 73(3)
8. Ashfaq Z, Mumtaz R, Rafay A, Zaidi SM H, Saleem H, Mumtaz S,Moerman I (2022) Embedded AI-based digi-healthcare. Appl Sci 12(1):519
9. Babar M, Khan MS, Din A, Ali F, Habib U, Kwak KS (2021) Intelligent computation offloading for IoT applications in scalable edge computing using artificial bee colony optimization. Complexity 2021:1–12
10. Brisimi TS, Chen R, Mela T, Olshevsky A, Paschalidis IC, Shi W (2018) Federated learning of predictive models from federated electronic health records. Int J Med Inf 112:59–67
11. Cheng W, Ou W, Yin X, Yan W, Liu D, Liu C (2020) A privacy-protection model for patients. Secur Commun Netw 2020:1–12
12. Dash S, Shakyawar SK, Sharma M, Kaushik S (2019) Big data in healthcare: management, analysis and future prospects. J Big Data 6(1):1–25
13. de la Torre L (2018) A guide to the california consumer privacy act of 2018. SSRN 3275571
14. Doudesis D, Lee KK, Boeddinghaus J, Bularga A, Ferry AV, Tuck C, Koechlin L (2023) Machine learning for diagnosis of myocardial infarction using cardiac troponin concentrations. Nat Med 1–10
15. Fang L, Liu X, Su X, Ye J, Dobson S, Hui P, Tarkoma S (2021) Bayesian inference federated learning for heart rate prediction. Paper presented at the wireless mobile communication and healthcare: 9th EAI international conference, mobihealth 2020, Virtual Event, 19 Nov 2020, Proceedings 9
16. Flora GD, Nayak MK (2019) A brief review of cardiovascular diseases, associated risk factors and current treatment regimes. Curr Pharm Des 25(38):4063–4084

17. Karaboga D (2010) Artificial bee colony algorithm. Scholarpedia 5(3):6915
18. Karaboga D, Akay B (2009) A comparative study of artificial bee colony algorithm. Appl Math Comput 214(1):108–132
19. Kelly KL, Lin PT, Basso C, Bois M, Buja LM, Cohle SD, Firchau D (2023) Sudden cardiac death in the young: a consensus statement on recommended practices for cardiac examination by pathologists from the Society for Cardiovascular Pathology. Cardiovasc Pathol 63:107497
20. Khan MA, Algarni F (2020) A healthcare monitoring system for the diagnosis of heart disease in the IoMT cloud environment using MSSO-ANFIS. IEEE Access 8:122259–122269
21. Li C, Hu X, Zhang L (2017) The IoT-based heart disease monitoring system for pervasive healthcare service. Procedia Comput Sci 112:2328–2334
22. Li JP, Haq AU, Din SU, Khan J, Khan A, Saboor A (2020) Heart disease identification method using machine learning classification in e-healthcare. IEEE Access 8:107562–107582
23. Ma Z, Zhao M, Cai X, Jia Z (2021) Fast-convergent federated learning with class-weighted aggregation. J Syst Arch 117:102125
24. Malinovskiy G, Kovalev D, Gasanov E, Condat L, Richtarik P (2020) From local SGD to local fixed-point methods for federated learning. Paper presented at the international conference on machine learning
25. Manimurugan S, Almutairi S, Aborokbah MM, Narmatha C, Ganesan S, Chilamkurti N, Almoamari H (2022) Two-stage classification model for the prediction of heart disease using IoMT and artificial intelligence. Sensors 22(2):476
26. Mateen Yaqoob M, Khurshid W, Liu L, Zulqarnain Arif S, Ali Khan I, Khalid O, Nawaz R (2022) Adaptive multi-cost routing protocol to enhance lifetime for wireless body area network. Comput Mater Contin 72(1):1089–1103
27. McMahan B, Moore E, Ramage D, Hampson S, y Arcas BA (2017) Communication-efficient learning of deep networks from decentralized data. Paper presented at the artificial intelligence and statistics
28. Nguyen HT, Sehwag V, Hosseinalipour S, Brinton CG, Chiang M, Poor HV (2020) Fast-convergent federated learning. IEEE J Sel Areas Commun 39(1):201–218
29. Nilsson A, Smith S, Ulm G, Gustavsson E, Jirstrand M (2018) A performance evaluation of federated learning algorithms. Paper presented at the proceedings of the second workshop on distributed infrastructures for deep learning
30. Sarmah SS (2020) An efficient IoT-based patient monitoring and heart disease prediction system using deep learning modified neural network. IEEE Access 8:135784–135797
31. Shuja J, Alanazi E, Alasmary W, Alashaikh A (2021) COVID-19 open source data sets: a comprehensive survey. Appl Intell 51:1296–1325
32. Voigt P, Von dem Bussche A (2017) The eu general data protection regulation (gdpr). A practical guide, 1st ed, vol 10(3152676). Springer international publishing, Cham, pp 10–5555
33. Wang H, Yurochkin M, Sun Y, Papailiopoulos D, Khazaeni Y (2020) Federated learning with matched averaging. arXiv:2002.06440
34. Yuan X, Chen J, Zhang K, Wu Y, Yang T (2021) A stable AI-based binary and multiple class heart disease prediction model for IoMT. IEEE Trans Industr Inf 18(3):2032–2040
35. Zaman SKU, Jehangiri AI, Maqsood T, Umar AI, Khan MA, Jhanjhi NZ, Masud M (2022) COME-UP: Computation offloading in mobile edge computing with LSTM based user direction prediction. Appl Sci 12(7):3312

Optimal Design of a Biomedical Amplifier for Minimum Offset Using a Modified ABC Algorithm

Naushad Manzoor Laskar, Swagata Devi, S Kumaravel, Mohammed Nazibul Hasan, Sagarika Choudhury, and Arjun Pandey

1 Introduction

In the development of modern biomedical systems, the design of an optimally performing front-end amplifier finds an important place in the overall design goals. Biomedical amplifiers are required to amplify the small amplitude (in μV ranges) and low frequency (few Hz to few KHz) biomedical systems, so as to enable a suitable pre-processing of these signals [1]. The key requirements for biomedical amplifiers include substantial gain, low power, and low noise [2] in addition to other important design parameters for optimal performance. Offset voltage stands out as a critical parameter in amplifier design, particularly in biomedical applications, where signal amplitudes are in the microvolt to lower millivolt range. The presence of a high offset voltage can significantly impact the accuracy and precision of the various

N. M. Laskar (✉) · S. Kumaravel · A. Pandey
School of Electronics Engineering, Vellore Institute of Technology, Vellore 632014, India
e-mail: naushadmanzoor.laskar@vit.ac.in

S. Kumaravel
e-mail: kumaravel.s@vit.ac.in

A. Pandey
e-mail: arjun.pandey2021@vitstudent.ac.in

S. Devi
Faculty of Engineering, Assam Down Town University, Sankar Madhab Path, Gandhi Nagar, Panikhaiti, Guwahati, Assam 781026, India

M. N. Hasan
Angstrom Laboratory, FENE Group, Uppsala University, Uppsala, Sweden
e-mail: mohammed.hasan@angstrom.uu.se

S. Choudhury
Tejas Networks Ltd., Bengaluru, Karnataka, India
e-mail: sagarikachoudhury36@gmail.com

K. Raza (ed.), *Solving with Bees*, Springer Tracts in Nature-Inspired Computing,
https://doi.org/10.1007/978-981-97-7344-2_7

above-mentioned design requirements. Among the various offset voltages, minimizing random offset poses a greater challenge as it results from inherent mismatches in transistors and other device parameters. While techniques like statistical modeling, advanced calibration methods, and precision-matched components can help mitigate its impact, complete elimination may prove difficult. Thus, tailoring the design to the specific requirements and constraints of biomedical amplifier is crucial to meet desired levels of accuracy and reliability.

In literature, various models such as Kinget [3] or Pelgrom's design model [4] have been explored, all sharing a common strategy to minimize random offset voltage. Pelgrom's model makes use of an inherent dependency of unpredictability in different transistor parameters such as threshold voltage, gate-to-source voltage, and drain current on the conventional shift in the area and placing among transistors that match. All these dependencies are modeled by Pelgrom's design transistor inconsistency model, which is discussed in detail in Sect. 2 of the manuscript. To minimize offset, be it Pelgrom's model or Kinget's or others, the process involves intricate analysis to enhance matching through advanced manufacturing processes, improved packaging materials, specialized layout techniques, or the incorporation of offset nullifying circuits. However, these approaches often result in increased area requirements. Another approach for minimizing random offset involves adjusting transistor dimensions, a method explored in the aforementioned models. However, the challenge lies in the fact that altering transistor dimensions not only impacts random offset but also influences other critical design specifications, as previously mentioned. Arbitrary sizing is therefore not feasible. Additionally, design specifications inherently involve trade-offs, and transistor sizing directly affects circuit area, which, in turn, impacts chip cost.

The main motivation behind the proposed work is to determine an optimal random offset voltage, considering these complex design trade-offs. Manual transistor sizing, being inefficient and not accounting for all constraints, prompts the need for optimization algorithms. These approaches attempt to meet design constraints like area constraints and specifications for the design with the goal to achieve the greatest random offset voltage. The use of optimization methods for random offset reduction in neural amplifiers was first proposed by Laskar et al. [5] utilizing a hybrid Whale Particle Swarm Optimization [6]. The procedure outlines how to simulate at random offset voltage for all type of circuit using the conventional Pelgrom's design Model and then compute a cost function that takes the offset voltage standard deviation into account. Hybrid whale particle swarm optimization (HWPSO) is then used to satisfy minimum design constraints and solve the derived Cost Function. Other major works in minimization of offset voltage using optimization algorithms, considering both systematic and random, include that by Sarkar et al. [7], Laskar et al. [8], etc. The present work uses a similar approach as in [5], however for biomedical applications and also a different amplifier topology namely the improved high swing self-biased folded cascode (HSSFC) with current scaling [9, 10]. The HSSFC has been reported to achieve improved performance in terms of noise performance, which is a major design requirement in biomedical amplifiers. Furthermore, the proposed work makes use of a high-performance modified Artificial Bee Colony (ABC) [11] which employs a selection strategy of a novel rate of change approach to modify the initial population and bee position based on average cost. The MABC-SS reportedly exercises arbitrary and opposition-based learning methods to initialize the population and revises a bee's position after surpassing an accurate number of trial caps. It has been reported to achieve promising outcomes as compared to various cutting-edge algorithms including other ABC variants [12, 13] and the algorithms reported in [5] such as the reputed Dragonfly Algorithm [14], Whale Optimization Algorithm [15], and Grasshopper Algorithm [16] for standard benchmark functions and this has prompted the use of this algorithm for such a complex real-time design optimization problem. Some other recently proposed algorithms of 2024 [17, 18] have also not been able to match MABC-SS in terms of exploration ability.

Extensive simulations have been conducted, and a relative analysis with different modern and advanced algorithms has been meticulously presented, leveraging statistical parameters [19]. The outcome demonstrates that MABC-SS surpasses the majority of the state-of-the-art algorithms [5, 12–14, 20] by attaining a minimum random offset while meeting the desired specifications and achieving a superior Friedmann's test rank. Further, the results have been validated in the Cadence circuit design tool with 180 nm technology and implementing Monte Carlo simulations. Notably, the reported outcomes exhibit keen alignment with the algorithmic outcomes, underscoring the effectiveness of the proposed approach. The remaining portions of the paper are arranged as follows: The problem is described in Sect. 2. Section 3 describes the MABC-SS algorithm. The article is summarized in Sect. 5 after the results are thoroughly discussed in Sect. 4.

2 Problem Description

To formulate the cost function, Pelgrom's design model is utilized to obtain the random offset voltage for an HSSFC biomedical amplifier, which is then minimized under several design constraints based on the design procedure which will be discussed in Sect. 2.2.

2.1 Pelgrom's Design Model

The intrinsic dependence of unpredictability in different transistor variables on the conventional shift in the area and placing among transistors that match are taken into account by Pelgrom's design transistor inconsistency model. These dissimilarities are considered to have a zero-mean normal distribution by Pelgrom's design model. Equation (1) can be used to express the model for any given process parameter, *P*.

$$\sigma^2(\Delta P) = S_P^2 D_P^2 + \frac{A_P^2}{WL} \tag{1}$$

Here, W, L give the dimensions, i.e. width/length of each transistor/component, D_x defines the spacing within each transistor, and $\sigma^2(\Delta P)$ is the variation of any particular parameter P difference among the elements. The threshold voltage (V_{TH}), the current factor (β), drain current (I_D), gate-to-source voltage, etc. are the most important parameters representing P that affect mismatch or random offset in circuits [21]. A_p^2 and S_p^2 are the area and the spacing proportional representation constant for every parameter 'P'. A modification to Eq. (1) was introduced by Kinget [3], who showed that the area outweighs the contribution of the spacing term. The altered formula is written as in Eq. (2). It is possible to relate the adjustment for gate-source voltage (V_{GS}) and the drain current (I_D) as functions of the two previously mentioned variables as shown in Eqs. (3) and (4), respectively.

$$\sigma^2(\Delta P) = \frac{A_P^2}{\mathrm{WL}} \tag{2}$$

$$\sigma^2(\Delta I_D) = \sigma^2(\Delta\beta) + \frac{g_m^2}{I_D^2}(\Delta V_{\mathrm{TH}}) \tag{3}$$

$$\sigma^2(\Delta V_{\mathrm{GS}}) = \sigma^2(\Delta V_{\mathrm{TH}}) + \frac{I_D^2}{g_m^2}\frac{\sigma^2(\Delta\beta)}{\beta} \tag{4}$$

where g_m is the ratio of output current to input voltage termed as transconductance and $\sigma^2(\Delta P)$ is the variance of any parameter P where P can be I_D, β, V_{GS}, and V_{TH}.

2.2 *Offset Voltage and Cost Function Formulation for HSSFC Amplifier*

The schematic for an HSSFC amplifier with current scaling is shown in Fig. 1 [9]. The differential pair in this case is represented by M0–M1, the cascode transistors are M4–M5, the current sink is M2–M3, the cascode current mirror load is created by M6–M9, and the self-biased high swing resistor is M12. For current scaling, the resistors R0 and R1 are indicated. The difference in drain current between devices perceived at the OTA's output is the main cause of the random offset voltage in this OTA. By separating the outcome of the circuit's transconductance (g_m), the variance is then related to the input. As shown in Eq. (3), the drain current change for any circuit is found using the corresponding $\mathrm{I_D}$ formula.

As stated in Eq. (4), the variation in $\mathrm{I_D}$ is regarded as a small signal that gets transmitted to the transistor's gate through g_m. Since transconductance efficiency (g_m/I_D) is generally high, it can be approximately expressed as follows: $\sigma^2(\Delta V_{\mathrm{GS}}) = \sigma^2(\Delta V_{\mathrm{TH}})$. The contributions from M0, M2, and M8 transistors that have an effect on the output are what determine the total output random offset. Equation (5) illustrates how to calculate the overall input random offset voltage by splitting this by the transconductance (g_{m0}). One of the major challenges in the work is to fulfill the design

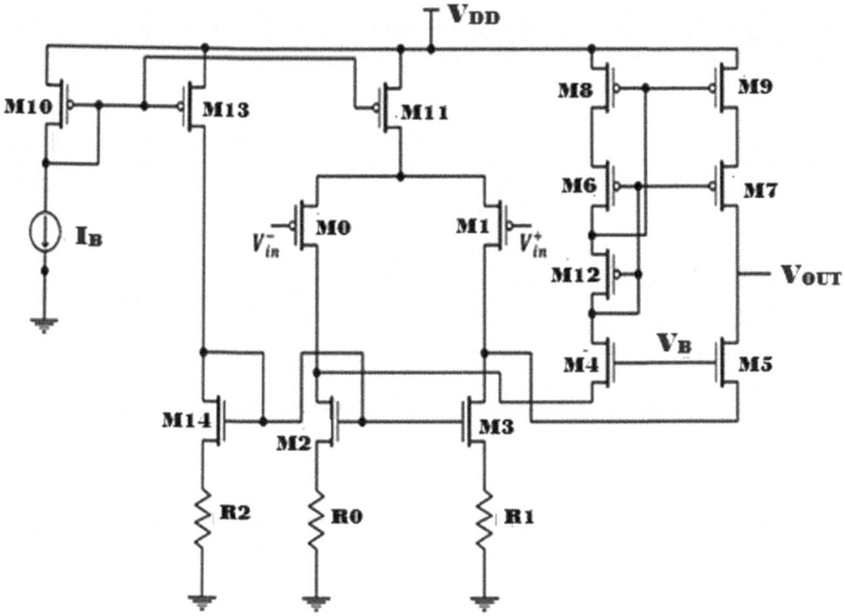

Fig.1 High swing self-bias folded Cascode (HSSFC) amplifier

requirements for the biomedical application shown in Table 1, while optimizing the desired objective function. This means that the offset voltage must be reduced while complying to the constraints defined by Eqs. (6)–(16). Furthermore, another challenge is that the optimal design specifications and the random offset voltage obtained by the algorithm has to be validated in the technological environment such as Cadence and there should not be much deviation in both the simulated values. The design vector for the optimization is outlined in Eq. (17).

$$V_{O,\text{HSSFC}} = 2\frac{A^2_{\text{VTH},P}}{W_0 L_0}\left[1 + \frac{2\mu_N A^2_{\text{VTH},N}}{\mu_P A^2_{\text{VTH},P}}\left(\frac{L_0}{L_2}\right)^2 + \left(\frac{L_0}{L_8}\right)^2\right] \quad (5)$$

$$I_{11} = (\text{S.R.})C_{\text{LOAD}} \quad (6)$$

$$\left(\frac{W}{L}\right)_3 = \frac{2I_3}{K_n V^2_{\text{DS3}}} = \left(\frac{W}{L}\right)_2 \quad (7)$$

$$\left(\frac{W}{L}\right)_5 = \frac{2I_5}{K_n V^2_{\text{DS5}}} = \left(\frac{W}{L}\right)_4 \quad (8)$$

where

$$V_{\mathrm{DS3}}(\mathrm{sat}) = V_{\mathrm{DS5}}(\mathrm{sat}) = \frac{V_{o(\min)}}{2}$$

$$\left(\frac{W}{L}\right)_9 = \frac{2I_9}{K_P V_{\mathrm{SD9}}^2} \text{ and } \left(\frac{W}{L}\right)_7 = \frac{2I_7}{K_P V_{\mathrm{SD7}}^2} \tag{9}$$

where

$$V_{\mathrm{SD7}}(\mathrm{sat}) = V_{\mathrm{SD9}}(\mathrm{sat}) = \frac{V_{\mathrm{DD}} - V_{o(\max)}}{2}$$

$$\left(\frac{W}{L}\right)_0 = \left(\frac{W}{L}\right)_1 = \frac{g_{m0}^2}{\frac{K_p 2I_2}{K+1}} = \frac{\mathrm{UGB}^2 C_L^2}{\frac{K_p 2I_2}{K+1}} \tag{10}$$

$$\left(\frac{W}{L}\right)_{11} = \frac{2I_{11}}{K_p\left[V_{\mathrm{DD}} - V_{\mathrm{IC(max)}} - \sqrt{\frac{2I_2}{K_p\left(\frac{W}{L}\right)_0}} - V_{\mathrm{tp}}\right]^2} \tag{11}$$

$$\left(\frac{W}{L}\right)_3 = \frac{2I_3}{K_n\left[V_{\mathrm{IC(min)}} + V_{\mathrm{tp}}\right]^2} = \left(\frac{W}{L}\right)_2 \tag{12}$$

$$A_{\mathrm{vd}} = g_{m1} g_{m5} r_{05}(r_{01}||r_{03}||R1)||(g_{m7} r_{07} r_{09}) \tag{13}$$

$$S_n(f) = \frac{8\mathrm{KT}}{g_{m0}^2}\left(\frac{g_{m0}}{3} + \frac{1}{R0} + \frac{2g_{m8}}{3} + \frac{2g_{m12}}{3}\right) \tag{14}$$

$$\mathrm{P.D.} = (V_{\mathrm{DD}})(I_{11} + I_9 + I_8) \tag{15}$$

$$\mathrm{Area} = \sum_{k=0}^{12} W_k L_k \tag{16}$$

$$X_{\mathrm{HSSFC}} = \left[S.R, A_{\mathrm{vd}}, \mathrm{UGB}, V_{\mathrm{IC(min)}}, V_{\mathrm{IC(max)}}, S_n(f), \mathrm{P.D.}\right] \tag{17}$$

3 Modified Artificial Bee Colony with Selection Strategy (MABC-SS)

Chaudhary [11] proposed the modified variant of ABC by modifying it with respect to the initial population formation and position update approach using new updating equations based on the rate of the change, which he termed as a selection strategy. The procedure grasps a given setback as input and iteratively undergoes phases involving employed bees, scout bees, onlooker bees, and strategy selection to produce

Table 1 Specifications of HSSFC for biomedical applications

Design criteria	Specifications
Slew rate, S.R. ($V/\mu s$)	≥ 10
Minimum input common mode range, $V_{IC,min}$ (V)	≥ -1.8
Maximum input common mode range, $V_{IC,max}$ (V)	≤ 1.8
Power dissipation, P.D. (μW)	≤ 10
Input noise, $S_n(f)$ (nV/$\sqrt{}$Hz)	≤ 400
Gain, A_{vd} (dB)	>40
Unity gain bandwidth, U.G.B (MHz)	≥ 2
Load capacitance, C_{LOAD} (pF)	≥ 1

the best solution. Output is generated when the iteration condition is met. While a random population initialization can prevent local minima entrapment, it may impede convergence speed [22]. The proposed algorithm utilizes Opposition-Based Learning (OBL) [23] along with a random approach for population initialization. It employs two strategies of selection, to address different problem types. Strategy 1 focuses on better exploitation for unimodal problems, speeding up the search process. Conversely, Strategy 2 enhances exploration to prevent getting stagnant at local optima in multidimensional problems. Employed and onlooker bees utilize these strategies, while scout bees remain unaffected. Each iteration decides whether to stick with the current strategy or switch. The proposed algorithm's logic unfolds as follows.

Population Initialization Phase

The algorithm initiates with the population initialization based on OBL, setting the groundwork for the optimization process.

Employed Bees Phase

The algorithm enters a loop across the employed bee's stage, where approaches are chosen dependent on the status value from the preceding iteration. Strategies guide the calculation of each employed bee's position in the swarm. Once each employed bee is processed, the onlooker bee phase begins.

Onlooker Bees Phase

Similar to the employed bees phase, the onlooker bee phase utilizes strategies determined by the status. Positions for each onlooker bee are calculated based on the chosen strategies. If scout bees are present, new solutions for each scout bee are computed.

Rate of Change and Status Update

The algorithm computes the degree of alteration and compares it with the former iteration to fix the status. If termination criteria are met, the best result becomes the output. Otherwise, the algorithm returns to the employed bee phase, continuing the iteration. This iterative process persists until the termination criteria are satisfied, ensuring the algorithm refines its solutions over successive loops.

Strategy Selection Technique

This is employed on the rate of the fitness change of any solution. The rate of change is the average of all fitness values in the bee swarm. For updating the fitness, two strategies, as defined in Eqs. (18) and (19), are used in employed bees and onlooker bees stages. For any iteration, if the current rate of fitness is less than the previous fitness, then employed bees will use Eq. (18) while Eq. (19) will be used by onlooker bees to update fitness. This is termed as Strategy 1. However, if the current rate of fitness exceeds the previous one, a combination of update equations based on four different scenarios is employed by the employed and onlooker bees. This is Strategy 2 and the update is made based on the complexity of the objective function by any of the four combinations.

a. Both Employed and Onlooker bees employ Eq. (18). This results in better exploration but slower convergence.
b. Employed uses Eq. (18) while Onlooker uses Eq. (19). This leads to exploration by employed and exploitation by onlookers.
c. The third scenario is the same as Strategy 1.
d. In the fourth scenario, both phases employ Eq. (19). But it leads to a local optimal solution despite having a faster convergence.

$$\begin{aligned}\text{New Fitness} = {} & \text{Current Fitness} \\ & + (\text{Random Number between} - 1 \text{ and } 1) \\ & * (\text{Current Fitness} - \text{Random Fitness in Swarm})\end{aligned} \tag{18}$$

$$\begin{aligned}\text{New Fitness} = {} & \text{Best Fitness} \\ & + (\text{Random Number between} - 1 \text{ and } 1) \\ & * (\text{Current Fitness} - \text{Random Fitness in Swarm})\end{aligned} \tag{19}$$

Reference [11] provides a more thorough explanation of the MABC-SS and its use with standard mathematical functions. Figure 2 shows the flowchart.

4 Results and Discussions

MABC-SS is applied to optimize the random offset voltage of the HSSFC Biomedical Amplifier as discussed in Sect. 2. MATLAB is used for the simulations, and other cutting-edge algorithms are benchmarked against MABC-SS to conduct a comprehensive comparative analysis [5, 12, 13, 15, 16]. Twenty separate runs, each with

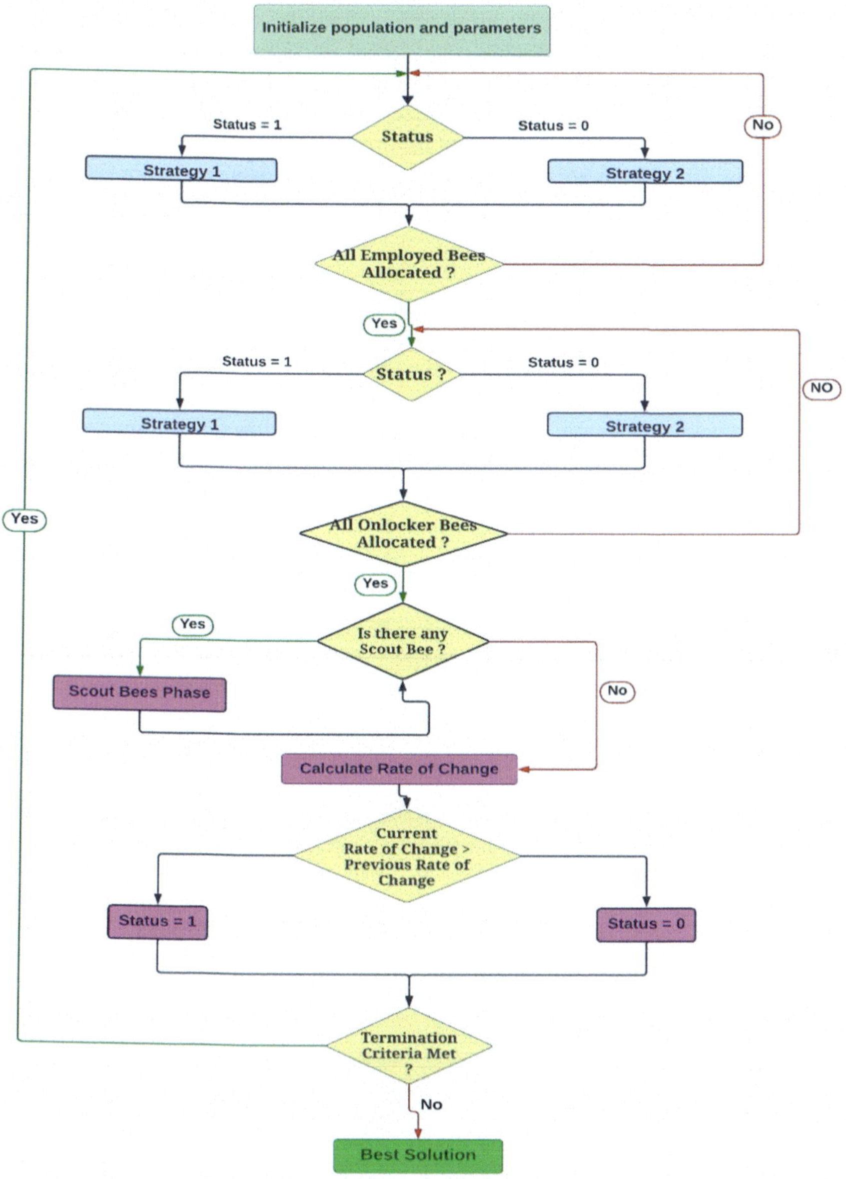

Fig. 2 Flowchart of MABC-SS [11]

the similar population dimension and design restrictions, are included in this analysis. Performance evaluation is conducted based on statistical analysis, including Friedman's rank and the Mann-Whitney U test, to discern the efficacy of MABC-SS in comparison to other algorithms. These tests provide a robust framework for assessing the consistency and reliability of the proposed optimization approach across multiple runs. The circuits are then practically redesigned using transistor sizes along with additional design variables that correspond to the requisite random offset voltage basically, the optimal values discovered during the algorithmic optimization process in order to validate the algorithmic results. This redesign is implemented in Cadence Virtuoso using 180 nm technology parameters, and necessary analyses are performed to confirm the alignment between algorithmic predictions and practical outcomes. This validation process further solidifies the reliability and practical applicability of the proposed optimization approach.

The optimization process involves minimizing the cost function outlined in Eq. (5) while adhering to the constraints defined by Eqs. (6)–(16) and Table 1. The algorithm selects an aggregate of 100 iterations, and more than twenty independent runs are carried out with different initial conditions to document the mean, standard deviation (SD), best, and worst values acquired by MABC-SS, as well as those from the other previously mentioned algorithms. The results, presented in Tables 2 and 3, highlight MABC-SS's superior performance in terms of both the best area and SD. This indicates the algorithm's robustness and consistency across different runs. The statistical significance of these results is confirmed through Friedman's rank test and Mann-Whitney's U test, as detailed in Tables 4 and 5, respectively. The p-values and U-values suggest that the outcomes are statistically important at a 5% level of implication.

Comparative analysis reveals that MABC-SS excels in designing an HSSFC Amplifier for biomedical applications, attaining a least (best) random offset voltage of 1.25 mV. The convergence plot in Fig. 3a illustrates that MABC-SS exhibits good convergence speed on par with other state-of-the-art algorithms in terms of iterations needed to achieve an optimum area. The search space begins with an initial optimum

Table 2 Comparative analysis with best results

Specifications	ABC	DA	OGJO	HWPSO	MABCSS
S.R. (V/μs)	174.27	176.3	176.26	175.08	173.36
A_{vd} (dB)	67.07	71.2	71.2	86.72	83.25
$S_n(f)$ (nV/$\sqrt{}$Hz)	13.37	13.48	13.48	13.52	13.50
P.D. (μW)	6.68	5.28	5.28	3.25	4.64
UGB (MHz)	5.82	4.07	4.07	4.47	5.32
$V_{IC,MIN}$ (V)	−0.038	−0.19	−0.19	−0.103	−0.155
$V_{IC,MAX}$ (V)	1.49	1.56	1.56	1.55	1.52
Area (m)^2	2.78e-10	6.66e-09	6.66e-09	4.86e-09	3.56e-09
$V_{O,HSSFC}$ (mV)	1.99	1.96	1.96	1.29	1.26

Table 3 Comparison of algorithms results (total independent runs = 20)

Optimum $V_{O,HSSFC}$ (mV)	ABC	DA	OGJO	HWPSO	MABCSS
Best	1.99	1.96	1.96	1.29	1.25
Worst	2.02	1.99	1.99	1.38	1.38
Mean	2.007	1.975	1.975	1.32	1.28
SD	0.88e-12	0.71e-12	0.71e-12	0.77e-12	0.86e-12

Table 4 Friedmann's statistical test results

Algorithm	Mean rank	Rank
ABC	5.862	5
DA	5.278	4
OGJO	4.952	3
HWPSO	4.863	2
MABCSS	4.778	1

Table 5 Mann-Whitney's statistical test results

MABCSS vs	p-value	U value (out of 100)	U value MABCSS
ABC	0.0000438	0	100
DA	0.0000107	3	97
OGJO	0.0000107	3	97
HWPSO	0.0000102	8	92

value of 190 mV and it explores the search space to achieve an optimum random offset. As seen from Fig. 3a, the optimum value decreases with every iteration and finally converges to an optimum value of 1.25 mV, after around 20 iterations, whereas the other state-of-the-art algorithms fail to the global minima obtained by MABC-SS and converge to a local optimal solution after approximately 25 iterations, thereby requiring a higher convergence time. The boxplot in Fig. 3b provides further insights, indicating that MABC-SS consistently performs well with optimum values closely clustered and a low standard deviation for more than 2000 function evaluations. The plot aligns with the data presented in Table 3, highlighting MABC-SS's reliability and efficiency in optimizing the random offset voltage for this specific case study.

As part of the validation process, the circuit is redesigned in Cadence Virtuoso using aspect ratios that match the optimal random offset voltage values found using MABC-SS. As presented in Table 2, the MABC-SS-based design achieves a gain of approximately 83.32 dB, coupled with a suitable cutoff frequency for biomedical applications. The input-referred noise is also within bounds measuring at 13.5 nV/$\sqrt{}$Hz, marginally improved as compared to HWPSO. The optimization process leads to specific values for W11, W9, thereby leading to improvement in I11,I9 and hence resulting in a positive impact on power dissipation, which is recorded at 4.64 μW.

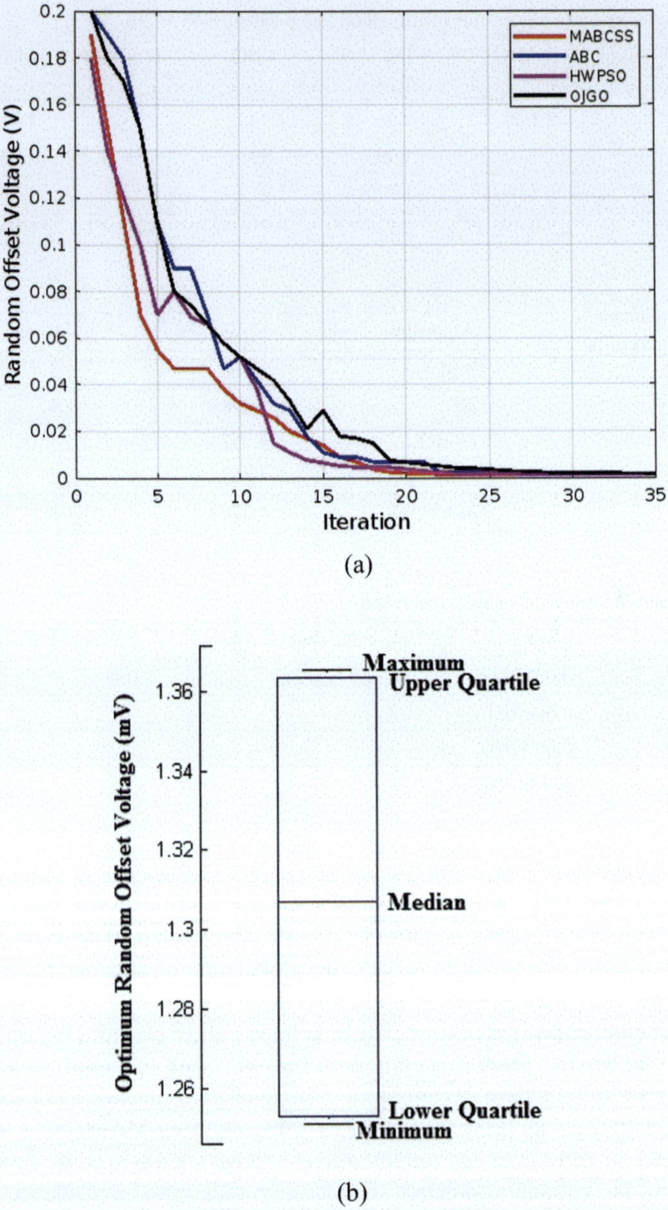

Fig. 3 **a** Convergence plot. **b** Box plot of MABC-SS

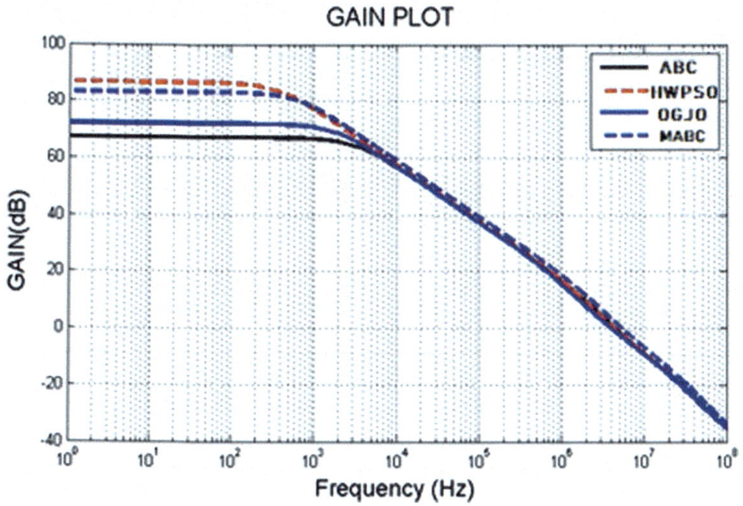

Fig. 4 Plot for gain validation in Virtuoso

The validation of MABC-SS results is meticulously performed by implementing a redesign of the HSSFC Amplifier in Cadence Virtuoso. This involves utilizing 180 nm technology parameters and conducting both pre-layout and Monte Carlo simulations. The results, illustrated in Figs. 4 and 5, closely align with the algorithmic predictions. The deviation observed in diverse design constraints, as specified in Figs. 4 and 5, is minimal. This level of difference can be accredited to parasitic properties besides ensuring high-order effects, as previously mentioned, which were neglected during the circuit modeling. This robust validation process substantiates the reliability and practical applicability of the MABC-SS algorithm in achieving optimal circuit design with minimized random offset voltage for the HSSFC Amplifier. The close agreement between algorithmic predictions and practical outcomes underscores the effectiveness of the proposed optimization approach.

5 Conclusions

An optimal sizing approach for HSSFC Amplifiers to minimize random offset voltage, specifically tailored for biomedical applications, is presented. The proposed methodology leverages the Modified Artificial Bee Colony with Selection Strategy (MABCSS) algorithm. The precision and accuracy of amplifier outputs can be greatly impacted by random offset voltage, which results from variations in process parameters among circuit devices. This is especially problematic in medical applications where the voltage inputs are of extremely low magnitude (lower microvolts to millivolts range). Pelgrom's design model is employed to obtain random offset voltage for the HSSFC architecture. MABCSS and other surveyed algorithms are then utilized to

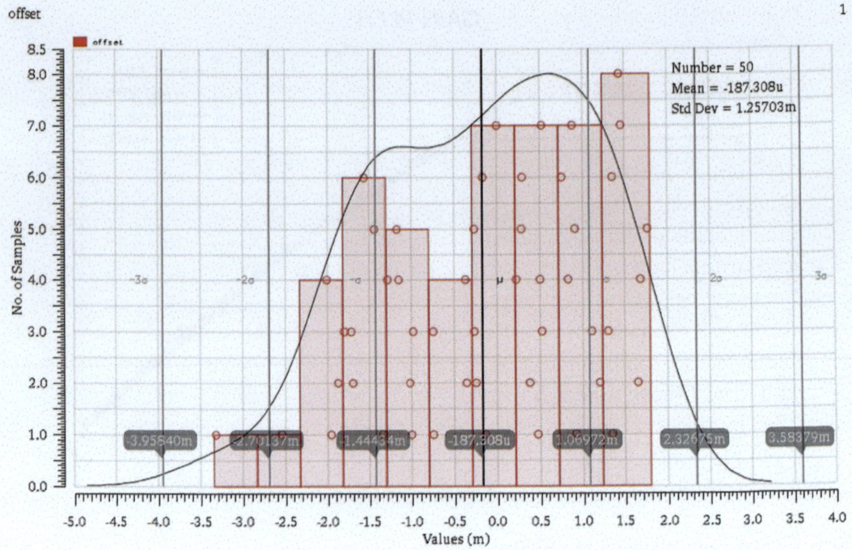

Fig. 5 Monte Carlo simulation for offset in Virtuoso

minimize offset voltages, considering various design constraints and specifications crucial for biomedical amplifiers.

MATLAB simulations are run for more than 25 separate runs under various initial conditions. A comparative analysis is carried out with a consistent population count of 50 and a maximum number of 100 iterations, capturing the worst, best, mean, and standard deviation for about 2500 function evaluations. Results display that when related to the current procedures for the circuit under consideration, the MABCSS algorithm quickly converges to an optimal random offset (best). In addition, the mean as well as the standard deviation outperforms the majority of cutting-edge algorithms, and the optimal values for the remaining design specifications are acceptable. When compared to other algorithms, statistical analyses confirm the consistency and dependability of the MABCSS algorithm. The computational results are validated using Cadence Virtuoso circuit simulations, where the ideal design parameter values are those that yield the best results. The minimal deviation is evident in the outcomes for a minimum random offset voltage, which closely align. The efficacy of the suggested strategy is confirmed by the strong correlation between algorithmic predictions and execution in other design specifications. The deviations can be assigned to certain higher-order properties, for example, body effect and velocity saturation. Thus, the MABCSS algorithm emerges as a powerful tool for minimizing random offset voltage, considering design trade-offs with low computational overhead. This contrasts with manual design approaches, which may struggle to account for all trade-offs simultaneously, potentially leading to higher random offset voltages and reduced design accuracy. The comprehensive validation process underscores the robustness and practical applicability of the proposed optimization approach.

References

1. Bharucha E, Sepehrian H, Gosselin B (2014) A survey of neural front end amplifiers and their requirements toward practical neural interfaces. J Low Power Electron Appl 4(4):268–291
2. Amaya JR, Perez AR, Restituto MD (2015) A low noise amplifier for neural spike recording interfaces. Sensors 15(10):25313–25335
3. Kinget P (2005) Device mismatch and tradeoffs in the design of analog circuits. IEEE J Solid-State Circ 40(6):1212–1224
4. Pelgrom MJM, Duinmaijer ACJ, Welbers APG (1989) Matching properties of MOS transistors. IEEE J Solid-State Circ 24(5):1433–1439
5. Laskar NM, Guha K, Paul PK, Baishnab KL (2021) Random offset minimization in low frequency front-end amplifiers using swarm intelligence based techniques. Evol Intel 14(3):1317–1335
6. Laskar NM, Guha K, Chatterjee I, Chanda S, Baishnab KL, Paul PK (2019) HWPSO: a new hybrid whale-particle swarm optimization algorithm and its application in electronic design optimization problems. Appl Intell 49:265–291
7. Sarkar P, Laskar NM, Nath S, Baishnab KL, Chanda S (2018) Offset voltage minimization based circuit sizing of CMOS operational amplifier using whale optimization algorithm. J Inf Optim Sci 39:83–98
8. Laskar NM, Guha K, Baishnab KL, Paul PK, Rao KS (2020) Optimizing the random offset voltage in two stage amplifier considering noise-power trade-off using HWPSO algorithm. In: Lecture notes electrical engineering: energy systems, drives and automations, vol 664, pp 349–359
9. Wattanapanitch W, Fee M, Sarpeshkar R (2007) An energy-efficient micropower neural recording amplifier. IEEE Trans Biomed Circuits Syst 1(2):136–147
10. Gupta M, Aggarwal B, Gupta AK (2013) A very high performance self-biased cascode current mirror for CMOS technology. Analog Integr Circ Sig Process 75(1):67–74
11. Chaudhary KC (2023) A modified version of the ABC algorithm and evaluation of its performance. Heliyon 9(5):e16086
12. Kang F, Li J, Ma Z (2011) Rosenbrock artificial bee colony algorithm for accurate global optimization of numerical functions. Inf Sci 181(16):3508–3531
13. Mohapatra S, Mohapatra P (2023) An improved golden jackal optimization algorithm using opposition-based learning for global optimization and engineering problems. Int J Comput Intell Syst 16(1):147
14. Mirjalili S (2016) Dragonfly algorithm: a new meta-heuristic optimization technique for solving single-objective, discrete, and multi- objective problems. Neural Comput Appl 27:1053–1073
15. Mirjalili S, Lewis A (2016) The whale optimization algorithm. Adv Eng Softw (Elsevier) 95:51–67
16. Saremi S, Mirjalili S, Lewis A (2017) Grasshopper optimisation algorithm: theory and application. Adv Eng Softw (Elsevier) 105:30–47
17. Gopi S, Mohapatra P (2024) Fast random opposition-based learning aquila optimization algorithm. Heliyon 10(4):e26187
18. El-Sayed, El-kenawy M, Khodadadi N, Mirjalili S, Abdelhamid A, Eid M et al (2024) Greylag goose optimization: nature-inspired optimization algorithm. Expert Syst Appl (Elsevier) 238:122147
19. Jakšić O, Jakšić Z, Guha K, Silva AG, Laskar NM (2023) Comparing artificial neural network algorithms for prediction of higher heating value for different types of biomass. Soft Comput 27(9):5933–5950

20. Das R, Das KN, Mallik S, Das S, Laskar NM, Nath S (2024) Novel self-adaptive multi-population quadratic approximation guided Jaya for solving real-parameter constrained optimization problems. Expert Syst Appl (Elsevier) 238:121898
21. Brito JPM, Bampi S, Klimach H (2007) A 4-bits trimmed CMOS bandgap reference with an improved matching modeling design. In: Circuits and systems IEEE international symposium, pp 1911–1914
22. Wang H, Wang W, Zhou X, Zhao J, Wang Y, Xiao S, Xu M (2021) Artificial bee colony algorithm based on knowledge fusion. Complex Intell Syst 7(3):1139–1152
23. Tizhoosh HR (2005) Opposition-based learning: a new scheme for machine intelligence. In: International conference on computational intelligence for modelling, control and automation and international conference on intelligent agents, web technologies and internet commerce (CIMCA-IAWTIC'06), vol 1, pp 695–701

Using Honeybees for Gene Expression Profiling: The Artificial Bee Colony Algorithm to Identify Robust Gene Biomarkers for Clinical Diagnosis

Sahar Qazi and Ashok Sharma

1 Introduction

The *Artificial Bee Colony* (ABC) algorithm, proposed by Karaboga [15], is a population-based optimization method inspired by the foraging behaviour of honeybees. In nature, bees exhibit efficient foraging strategies to locate food sources, which serve as the foundation for the ABC algorithm's exploration and exploitation mechanisms. Through the division of bees into employed, onlooker, and scout roles, the ABC algorithm iteratively searches for optimal solutions within a defined search space. This nature-inspired metaheuristic has garnered significant attention in the optimization community due to its simplicity, effectiveness, and versatility in solving various optimization problems across different domains [16].

Gene expression (GE) profiling plays a pivotal role in elucidating the intricate mechanisms underlying various biological processes, including development, differentiation, disease progression, and response to treatment. By quantifying the expression levels of thousands of genes simultaneously, this powerful technique provides valuable insights into the molecular signatures associated with different cellular states or disease conditions. Furthermore, gene expression profiling enables the identification of prognostic markers for disease diagnosis, prognosis, and therapeutic stratification, facilitating personalized medicine approaches [19]. Moreover, it aids in unravelling the complex regulatory networks governing gene expression, thus deepening our understanding of fundamental biological processes. With advancements in high-throughput sequencing technologies and bioinformatics analysis tools, gene expression profiling has become increasingly accessible and affordable, driving its widespread adoption across diverse fields of biomedical research and clinical practice [11].

S. Qazi (✉) · A. Sharma
Department of Biochemistry, All India Institute of Medical Sciences (AIIMS), New Delhi, India
e-mail: sqaziepiaiims@gmail.com

K. Raza (ed.), *Solving with Bees*, Springer Tracts in Nature-Inspired Computing,
https://doi.org/10.1007/978-981-97-7344-2_8

Honeybees have emerged as a valuable model organism for studying gene expression due to their sophisticated social behaviour, well-defined caste system, and extensive molecular tools available for research. The division of labour within honeybee colonies, with distinct roles assigned to workers, drones, and the queen, provides an excellent system for investigating the regulatory mechanisms underlying behavioural and physiological traits [25]. Gene expression studies in honeybees have elucidated key molecular pathways involved in various processes, including foraging behaviour, caste determination, immunity, and communication. Moreover, the sequencing of the honeybee genome has facilitated genome-wide expression analyses, enabling researchers to explore gene expression dynamics across different developmental stages, castes, and environmental conditions. By leveraging the honeybee model, researchers gain insights into conserved molecular mechanisms underlying social behaviour and complex traits across diverse organisms, including humans [26].

2 Gene Expression Profiling Techniques

Gene expression profiling techniques encompass a range of methodologies aimed at quantifying the activity of genes within a biological sample. These techniques provide valuable insights into the transcriptional dynamics underlying cellular processes, development, diseases, and responses to various stimuli. Microarray technology was one of the pioneering methods for gene expression profiling, allowing simultaneous measurement of thousands of genes' expression levels [24]. However, with advancements in sequencing technologies, RNA sequencing (RNA-seq) has become the gold standard due to its higher sensitivity, wider dynamic range, and ability to detect novel transcripts and alternative splicing events [13, 35, 36]. Additionally, single-cell RNA-seq enables gene expression profiling at the single-cell level, offering unprecedented resolution to dissect cellular heterogeneity within complex tissues and organisms [23]. Other emerging techniques, such as chromatin immunoprecipitation followed by sequencing (ChIP-seq), assay for transposase-accessible chromatin using sequencing (ATAC-seq), and spatial transcriptomics, provide complementary information on chromatin accessibility, epigenetic modifications, and spatial organization of gene expression, respectively [31]. Integrating these diverse profiling techniques enables comprehensive characterization of gene regulatory networks, paving the way for a deeper understanding of biological systems in health and disease.

Gene expression profiling encompasses various methods and technologies for quantifying the abundance of transcripts within a biological sample. These techniques have evolved over time, each offering unique advantages in terms of sensitivity, resolution, and throughput. The common gene profiling techniques that are used today are as follows:

(a) ***Microarray Technology***: Microarrays allow simultaneous measurement of the expression levels of thousands of genes. They rely on the hybridization of labelled cDNA or RNA to complementary probes immobilized on a solid support. Microarrays have been widely used for large-scale gene expression studies, enabling researchers to explore transcriptional changes associated with different biological conditions [24, 28].

(b) ***RNA Sequencing (RNA-seq)***: RNA-seq has revolutionized gene expression profiling by providing digital readouts of transcript abundance. This technology involves sequencing cDNA synthesized from RNA molecules present in the sample. RNA-seq offers higher sensitivity, wider dynamic range, and the ability to detect novel transcripts and alternative splicing events compared to microarrays [32, 35].

(c) ***Single-cell RNA Sequencing (scRNA-seq)***: scRNA-seq enables gene expression profiling at the single-cell level, providing insights into cellular heterogeneity within complex tissues and organisms. This technology has contributed to uncovering rare cell populations, identifying cell types, and dissecting transcriptional programs underlying cell differentiation and function [23, 31].

(d) ***Chromatin Immunoprecipitation Sequencing (ChIP-seq)***: ChIP-seq is used to investigate protein-DNA interactions, such as transcription factor binding and histone modifications. It involves immunoprecipitation of protein-DNA complexes followed by high-throughput sequencing of the isolated DNA fragments, allowing genome-wide mapping of transcription factor binding sites and histone modification patterns [21].

(e) ***Assay for Transposase-Accessible Chromatin Sequencing (ATAC-seq)***: ATAC-seq is a technique used to assess chromatin accessibility, providing insights into regions of open chromatin associated with active transcriptional regulatory elements. It involves the use of a hyperactive Tn5 transposase to insert sequencing adapters into accessible regions of chromatin, followed by high-throughput sequencing [5].

(f) ***Quantitative PCR (qPCR)***: qPCR is a widely used method for validating gene expression data obtained from high-throughput techniques. It provides accurate and quantitative measurement of the expression levels of specific target genes [6].

(g) ***Spatial Transcriptomics***: This emerging technology allows the visualization of gene expression within the context of tissue architecture, providing spatial information about the location of transcripts in situ [30].

3 Challenges and Limitations of Gene Expression Profiling Techniques

Gene expression profiling techniques face several challenges and limitations that impact their accuracy, reproducibility, and applicability. These challenges include the high cost and complexity associated with some methods, such as RNA-seq and single-cell RNA-seq, which require specialized equipment and expertise for data analysis and interpretation [35]. Additionally, issues related to sample preparation, RNA quality, and handling can introduce biases and artefacts into the data. Data analysis presents another significant challenge, with variability in pre-processing and normalization methods affecting the reproducibility and reliability of results [33]. Furthermore, detection limits and dynamic range limitations may lead to incomplete representation of the transcriptome, particularly for low-abundance transcripts or highly expressed genes. Addressing these challenges will be essential for advancing gene expression profiling techniques and maximizing their utility in biological research and clinical applications [9, 12].

4 The ABC Algorithm and Its Applications in Gene Expression Profiling

The ABC algorithm, inspired by the foraging behaviour of honeybees, has gained prominence as a powerful optimization technique. The ABC algorithm models the collective intelligence of bee colonies, employing employed, onlooker, and scout bees to explore solution spaces iteratively. While primarily applied in optimization problems, the versatility of the ABC algorithm extends to various domains, including bioinformatics. In gene expression profiling, where the analysis of transcriptomic data plays a crucial role, the ABC algorithm demonstrates potential in optimizing the identification of differentially expressed genes and refining the clustering of expression patterns. This application leverages the algorithm's ability to efficiently navigate complex solution spaces, making it a promising tool for enhancing the accuracy and efficiency of gene expression analysis. As researchers seek innovative computational approaches to extract meaningful insights from high-dimensional genomic data, the utilization of the ABC algorithm in gene expression profiling showcases its adaptability and effectiveness in addressing challenges within this domain [29].

One example of applying the Artificial Bee Colony algorithm in next-generation sequencing is in the optimization of read alignment parameters. NGS data analysis often involves aligning millions of short sequencing reads to a reference genome, a computationally intensive task. The alignment process requires selecting appropriate parameters, such as the alignment algorithm, seed length, and mismatch allowance, which significantly impact alignment accuracy and computational efficiency. The ABC algorithm can be utilized to optimize these alignment parameters by iteratively exploring the parameter space to maximize alignment accuracy and minimize computational resources [17]. Employing the ABC algorithm in this context involves defining a set of candidate solutions (parameter configurations), evaluating their performance based on alignment quality metrics (e.g., mapping rate, alignment score), and iteratively updating the solutions guided by the principles of bee colony foraging behaviour. During each iteration, employed bees explore the parameter space by adjusting the alignment parameters (seed length, mismatch allowance) within predefined ranges. Onlooker bees evaluate the quality of these solutions based on alignment metrics, and scout bees introduce new parameter configurations to diversify the search space. Through this iterative process of exploration and exploitation, the ABC algorithm converges towards optimal alignment parameter values that maximize alignment accuracy and minimize computational resources. It helps in optimizing the read alignment parameters in next-generation sequencing analyses aiding researchers to improve the accuracy of downstream analyses, such as variant calling and transcript quantification, thus, ultimately improving the biological insights [1].

5 Honeybees in Gene Expression Studies

The rationale for using the ABC algorithm in cancer prediction lies in its ability to effectively handle optimization problems, including feature selection and classification tasks based on gene expression profiles. The ABC algorithm is a population-based optimization algorithm inspired by the foraging behaviour of honeybees, which makes it suitable for exploring and exploiting solutions in a complex search space. In the context of cancer prediction, gene expression profiles provide a high-dimensional dataset, and selecting relevant features for accurate classification is crucial [11]. The algorithm's main rationale is grounded in the idea that ABC, with its swarm intelligence-inspired approach, can efficiently identify a subset of genes that are indicative of cancer types, leading to more accurate and reliable predictions. The optimization capability of ABC is particularly valuable in dealing with the high dimensionality and noisy nature of gene expression data, making it a promising tool for cancer prediction and classification tasks.

Many studies have investigated the application of the ABC algorithm in cancer identification through gene expression profiles. Dhanabal and Arunachalam [8] explored ABC's effectiveness in cancer classification, particularly in optimizing feature selection to improve diagnostic accuracy. Additionally, Rajesh and Radha [22] proposed an enhanced version of the ABC algorithm tailored for gene selection from microarray data, specifically for cancer classification tasks. These studies, among others, highlight the potential of the ABC algorithm in effectively analysing gene expression profiles for cancer identification and classification [8, 22]. These studies have highlighted the algorithm's ability to navigate through the vast search space of gene expression data and identify relevant features that contribute to improved predictive performance in cancer diagnosis.

The ABC algorithm plays a significant role in biomarker identification, particularly in the field of bioinformatics and biomedical research. Biomarkers are biological markers that can be objectively measured and evaluated as indicators of normal biological processes, pathogenic processes, or pharmacological responses to therapeutic interventions. Identifying biomarkers is crucial for various applications, including disease diagnosis, prognosis, and personalized medicine. One of the primary challenges in biomarker identification is selecting relevant features (genes, proteins, metabolites, etc.) from high-dimensional datasets such as gene expression profiles or omics data. ABC, with its optimization capabilities, can efficiently explore the feature space and select subsets of features that are most discriminative for distinguishing between different biological conditions (diseased vs. healthy samples) [20].

Prognostic marker identification also involves optimizing certain criteria, such as maximizing classification accuracy, minimizing misclassification error, or maximizing the predictive power of the selected biomarkers. ABC's ability to navigate complex search spaces and balance exploration and exploitation makes it well-suited for optimizing these criteria, leading to the discovery of robust and informative biomarkers. Additionally, ABC can be integrated with various machine learning models, such as support vector machines (SVMs), random forests, or neural networks, to further enhance biomarker identification. By optimizing feature selection in conjunction with model parameters, ABC can improve the overall performance of biomarker-based predictive models. The algorithm is computationally efficient and scalable, making it suitable for analysing large-scale datasets commonly encountered in biomarker identification studies. It can handle thousands to millions of features and samples efficiently, facilitating the discovery of biomarkers across diverse biological contexts [35].

Many hybrid algorithms, such as Co-ABC, mRMR-ABC, SVM-ABC, and others, have been proposed to enhance the diagnosis of various cancers, including breast cancer, leukaemia, colon cancer, lung cancer, and more, by leveraging microarray and high-throughput sequencing-based gene expression profiling data. For instance, Alshamlan et al. [3] introduced Co-ABC, integrating the Artificial Bee Colony (ABC) algorithm with cooperative co-evolution techniques, while Alshamlan et al. [4] developed mRMR-ABC, combining ABC with Minimum Redundancy Maximum Relevance feature selection. Additionally, SVM-ABC [18], which integrates ABC with SVMs, has been explored for cancer classification tasks. These studies demonstrate the effectiveness of hybrid approaches in optimizing feature selection, improving classification accuracy, and ultimately enhancing cancer diagnosis [4]. *Artificial Bee Colony based on Dominance* (ABCD) algorithm is a hybrid model that uses the potential of AHP-based gene filtering, SVM, and ABC algorithm for hybrid gene selection for cancer research and prognostic marker discovery [7]. Table 1 summarizes the relevant ABC algorithms that are popularly used for gene expression profiling.

The ABC algorithm serves as a powerful tool for biomarker identification by enabling efficient feature selection, optimizing criteria, handling noisy data, integrating with machine learning models, and offering scalability and efficiency in analysis. Its application in biomarker discovery contributes to advancing our understanding of diseases, improving diagnostic and prognostic accuracy, and facilitating the development of personalized therapeutic strategies.

Table 1 Commonly used ABC and hybridized algorithms for gene expression profiling

Name	Description	Advantages	References
mRMR-ABC	Minimum redundancy maximum relevance-ABC Algorithm	The algorithm achieves better classification performance with a handful number of genes	[4]
Co-ABC	Correlation artificial bee colony algorithm	Co-ABC algorithm works well with a small number of predictive genes	[3]
SVM-ABC	Support vector machine-artificial bee colony algorithm	The algorithm has less rigid parameters, faster convergence speed, and stronger searching ability than other algorithms available	[18]
Deep learning-based ABC	Deep learning-based artificial bee colony algorithm	The algorithm is a wrapper-based feature selection system that allows to select fewest genes with greater prediction accuracy	[2]
Mo-ABC	Modified-artificial bee colony (ABC) utilizing fuzzy C	Mo-ABC provides a higher grouping precision and takes fewer bunching periods after contrasted and existing calculations	[14]
ABCD	Artificial bee colony based on dominance	A hybrid method that starts with the filtering of the genes of the dataset to prevent the high-dimensionality problem of gene expression datasets	[7]
ACO-ABC	Ant colony optimization-artificial bee colony algorithm	More accurate classification results for a smaller number of selected genes	[20]
AMC-ABC	Archive-based multi-criteria artificial bee colony algorithm	It provides fast convergence compared to the basic ABC with a higher coverage for the problems with both low and high difficulty levels	[27]

6 Current Challenges and Future Prospects in Clinical Diagnosis

The deployment of the ABC algorithm in the clinical diagnosis of cancers presents both challenges and future prospects. One challenge is the interpretability of the results generated by the ABC algorithm. While ABC can effectively identify biomarkers and optimize classification models, understanding the biological significance of the selected features and the decision-making process of the algorithm remains a challenge. Employment of diverse data types, such as genomic, proteomic, and clinical data, poses a challenge. ABC algorithms need to handle heterogeneous data sources to provide comprehensive insights for clinical diagnosis. Evaluation of the performance and generalizability of ABC-based models in clinical settings is crucial. Large-scale clinical trials and rigorous validation studies are needed to demonstrate the effectiveness and reliability of ABC algorithms in real-world applications [10].

With the current challenges in hand, there are a few promising prospects too! The ABC algorithms hold promise for advancing personalized medicine by identifying patient-specific biomarkers and treatment strategies. Future research may focus on tailoring ABC-based diagnostic tools to individual patients for more precise and effective cancer management. Moreover, integrating ABC with other artificial intelligence (AI) techniques such as deep learning could enhance its capabilities in analysing complex cancer datasets. Hybrid approaches that combine ABC with deep learning models may offer improved accuracy and scalability for cancer diagnosis [37]. There is potential for deploying ABC algorithms in real-time diagnosis systems, enabling rapid and accurate cancer detection and classification. This could facilitate early intervention and improve patient outcomes [34].

7 Conclusion

In conclusion, the ABC algorithm offers a wide range of potential benefits for gene expression profiling for biomarker discovery in cancer. Through its capacity to efficiently optimize feature selection processes, ABC facilitates the identification of key molecular signatures indicative of different cancer types. However, challenges persist, including the need for enhanced interpretability, scalability to larger datasets, and robustness to noise. Future directions may involve integrating ABC with multiomics data and advanced machine learning techniques to unlock its full potential in personalized cancer diagnostics and therapeutics. As research continues, addressing these challenges and harnessing the opportunities afforded by ABC will undoubtedly propel biomarker discovery forward in the fight against cancer.

References

1. Ali MZ, Thakur M (2019) ABC optimized neural network for gene expression data classification. In: International conference on intelligent communication and computational techniques. Springer, Singapore, pp 595–603
2. Almarzouki HZ (2022) Deep-learning-based cancer profiles classification using gene expression data profile. J Healthc Eng 2022:1–13. https://doi.org/10.1155/2022/4715998
3. Alshamlan HM (2018) Co-ABC: correlation artificial bee colony algorithm for biomarker gene discovery using gene expression profile. Saudi J Biol Sci 25(5):895–903. https://doi.org/10.1016/j.sjbs.2017.12.012
4. Alshamlan H, Badr G, Alohali Y (2015) MRMR-ABC: a hybrid gene selection algorithm for cancer classification using microarray gene expression profiling. Biomed Res Int 2015:1–15. https://doi.org/10.1155/2015/604910
5. Buenrostro JD, Giresi PG, Zaba LC, Chang HY, Greenleaf WJ (2013) Transposition of native chromatin for fast and sensitive epigenomic profiling of open chromatin, DNA-binding proteins and nucleosome position. Nat Methods 10(12):1213–1218
6. Bustin SA, Benes V, Garson JA, Hellemans J, Huggett J, Kubista M, Wittwer CT (2009) The MIQE guidelines: minimum information for publication of quantitative real-time PCR experiments. Clin Chem 55(4):611–622
7. Coleto-Alcudia V, Vega-Rodríguez MA (2020) Artificial bee colony algorithm based on dominance (ABCD) for a hybrid gene selection method. Knowl-Based Syst 205:106323–106323. https://doi.org/10.1016/j.knosys.2020.106323
8. Dhanabal D, Arunachalam VP (2012) Cancer classification using gene expression data with artificial bee colony algorithm. Int J Comput Appl 52(3)
9. Dillies MA, Rau A, Aubert J, Hennequet-Antier C, Jeanmougin M, Servant N, Schaeffer B (2013) A comprehensive evaluation of normalization methods for Illumina high-throughput RNA sequencing data analysis. Brief Bioinform 14(6):671–683
10. Gao Y, Zhang Y, Chen M, Zhao Q, Chen X (2020) A review of artificial bee colony (ABC) algorithm and its applications to medical data analysis. Comput Math Methods Med 2020:7809163
11. Goh WWB, Wang W, Wong L (2017) Why batch effects matter in omics data, and how to avoid them. Trends Biotechnol 35(6):498–507
12. Imbeaud S, Graudens E, Boulanger V, Barlet X, Zaborski P, Eveno E, De Reyniès A (2005) Towards standardization of RNA quality assessment using user-independent classifiers of microcapillary electrophoresis traces. Nucleic Acids Res 33(6):e56–e56
13. Jabeen A, Ahmad N, Raza K (2018) Machine learning-based state-of-the-art methods for the classification of RNA-Seq data. In: Dey N, Ashour A, Borra S (eds) Classification in BioApps. Lecture notes in computational vision and biomechanics, vol 26. Springer, pp 133–172. https://doi.org/10.1007/978-3-319-65981-7_6
14. Jackson (2020) Gene expression using artificial bee colony besides fuzzy C means and NFDA. Int J Psychosoc Rehabil 24(3):2028–2036. https://www.psychosocial.com/article/PR200949/11264
15. Karaboga D (2005) An idea based on honey bee swarm for numerical optimization. Technical Report-TR06, Erciyes University, Engineering Faculty, Computer Engineering Department
16. Karaboga D, Basturk B (2007) A powerful and efficient algorithm for numerical function optimization: artificial bee colony (ABC) algorithm. J Global Optim 39(3):459–471
17. Li H, Durbin R (2009) Fast and accurate short read alignment with Burrows-Wheeler transform. Bioinformatics 25(14):1754–1760
18. Lu J, Liao X, Li S, Ouyang H, Chen K, Huang B (2019) An effective ABC-SVM approach for surface roughness prediction in manufacturing processes. In: Complexity, New York, N.Y., pp 1–13. https://doi.org/10.1155/2019/3094670
19. Luo W, Friedman MS, Shedden K, Hankenson KD, Woolf PJ (2009) GAGE: generally applicable gene set enrichment for pathway analysis. BMC Bioinform 10(1):161

20. Moosa JM, Shakur R, Kaykobad M, Rahman MS (2016) Gene selection for cancer classification with the help of bees. BMC Med Genomics 9(S2). https://doi.org/10.1186/s12920-016-0204-7
21. Park PJ (2009) ChIP-seq: advantages and challenges of a maturing technology. Nat Rev Genet 10(10):669–680
22. Rajesh R, Radha P (2012) An improved artificial bee colony algorithm for gene selection from microarray data for cancer classification. Procedia Eng 38:2434–2441
23. Raza K (2024) Machine learning in single-cell RNA-seq data analysis. In: SpringerBriefs in computational intelligence. Springer, https://doi.org/10.1007/978-981-97-6703-8
24. Raza K (2016) Analysis of microarray data using artificial intelligence based techniques. In: Handbook of research on computational intelligence applications in bioinformatics. IGI Global, USA, pp 216–239. https://doi.org/10.4018/978-1-5225-0427-6.ch011
25. Robinson GE, Vargo EL (1997) Juvenile hormone in adult eusocial hymenoptera: gonadotropin and behavioral pacemaker. Arch Insect Biochem Physiol: Publ Collab Entomol Soc Am 35(4):559–583
26. Sadd BM, Barribeau SM (2015) Heterogeneity in infection outcome: lessons from a bumblebee-trypanosome system. Parasite Immunol 37(7):334–336
27. Sahin O, Akay B, Karaboga D (2021) Archive-based multi-criteria artificial bee colony algorithm for whole test suite generation. Eng Sci Technol Int J 24(3):806–817. https://doi.org/10.1016/j.jestch.2020.12.011
28. Schena M, Shalon D, Davis RW, Brown PO (1995) Quantitative monitoring of gene expression patterns with a complementary DNA microarray. Science 270(5235):467–470
29. Srinivas M, Patnaik LM (2012) Particle swarm optimization: a survey. Int J Swarm Intell Res 3(1):1–28
30. Ståhl PL, Salmén F, Vickovic S, Lundmark A, Navarro JF, Magnusson J, Frisén J (2016) Visualization and analysis of gene expression in tissue sections by spatial transcriptomics. Science 353(6294):78–82
31. Svensson V, Pachter L (2018) RNA velocity: molecular kinetics from single-cell RNA-seq. Mol Cell 72(1):7–9
32. Tang F, Barbacioru C, Wang Y, Nordman E, Lee C, Xu N, Surani MA (2009) MRNA-Seq whole-transcriptome analysis of a single cell. Nat Methods 6(5):377–382
33. Tarazona S, García-Alcalde F, Dopazo J, Ferrer A, Conesa A (2011) Differential expression in RNA-seq: a matter of depth. Genome Res 21(12):2213–2223
34. Teo J, Sen S (2013) A review on bilevel evolutionary algorithms for gene selection in cancer classification. Brief Bioinform 14(4):397–410
35. Wang Z, Gerstein M (2009) RNA-Seq: a revolutionary tool for transcriptomics. Nat Rev Genet 10(1):57–63
36. Yang L, Bhat AM, Qazi S, Raza K (2023) DLC1 as Druggable target for specific subsets of gastric cancer: an RNA-seq-based study. Medicina 59(3):514. https://doi.org/10.3390/medicina59030514
37. Zhang J et al (2012) A novel hybrid approach based on artificial bee colony and cooperative co-evolution for gene selection in microarray data for cancer classification. Knowl-Based Syst 36:259–265

Smart Diagnostics for Diabetic Retinopathy: Integrating Artificial Bee Colony Algorithms into Medical Image Analysis

R. S. M. Lakshmi Patibandla, B. Tarakeswara Rao, and M. Ramakrishna Murthy

1 Introduction

Millions of people worldwide suffer from the severe consequence of diabetes called diabetic retinopathy (DR), which has a protracted and dire outlook. It deprives people of their sight and independence, is the primary cause of blindness in working-age adults, and poses a danger to the health of eyesight worldwide. The true cost of mental illness is enormous in terms of both societal and economic consequences and goes beyond personal misery, necessitating immediate attention and intervention. 4.2 million of the 34 million persons who now have DR have vision-threatening types of the disease. According to estimates from the World Health Organization, this deceptive disease has already rendered 2.85 million people blind. These numbers are predicted to increase much more because of the worrying global increase in diabetes prevalence. It's also concerning to consider the financial cost of DR [1, 2]. The annual expenditures of screening, treatment, and vision rehabilitation can amount to billions of dollars, placing a heavy burden on healthcare systems. The social consequences are disastrous, even before one considers the financial ones. DR-related vision loss has a major influence on people, families, and communities since it can result in employment loss, a decline in quality of life, and a reliance on others. The good news is that, with early detection and intervention being crucial, DR is primarily preventable and treatable. Regretfully, DR frequently advances silently and without

R. S. M. Lakshmi Patibandla (✉)
Koneru Lakshmaiah Education Foundation, Guntur, Andhra Pradesh, India
e-mail: patibandla.lakshmi@gmail.com

B. Tarakeswara Rao
Kallam Haranadhareddy Institute of Technology, Guntur, Andhra Pradesh, India

M. Ramakrishna Murthy
Anil Neerukonda Institute of Technology and Sciences, Visakhapatnam, Andhra Pradesh, India

K. Raza (ed.), *Solving with Bees*, Springer Tracts in Nature-Inspired Computing,
https://doi.org/10.1007/978-981-97-7344-2_9

any early signs, which means that by the time vision loss manifests, the disease may already be advanced. This emphasizes how crucial it is to get a diagnosis quickly and accurately, as doing so can stop 95% of DR-related vision loss. For early detection, routine eye exams are essential. Comprehensive eye exams should be performed on all diabetics at least once a year; screenings that are more regular are advised for those who pose a higher risk. Retinal photography and telemedicine are examples of accessible screening methods that are essential for reaching underprivileged communities and guaranteeing early diagnostic availability. Additionally, for a proper diagnosis and treatment of DR, there must be a enough number of qualified ophthalmologists and eye care professionals [3].

DR impact requires more than just diagnosis; a multifaceted strategy is needed to combat it. Maintaining blood sugar control is crucial for managing diabetes because it lowers the chance of developing diabetic kidney disease and slows its progression. Campaigns to raise public awareness about diabetes-related disorders, their dangers, and the value of early detection are essential in educating both the general public and those with diabetes. Long-term DR management requires better diagnosis, therapies, and preventive strategies, all of which can only be achieved by funding research.

While not an easy one, diabetic retinopathy is a serious threat to the world's vision health. We can reverse the catastrophic effects of this disease by investing in healthcare professionals, prioritizing accessible screening programs, promoting efficient diabetes management, and funding ongoing research. The secret to protecting the priceless gift of sight for millions of people worldwide and guaranteeing a better future for all is early detection, precise diagnosis, and a comprehensive approach.

2 Artificial Bee Colony Algorithms

The Artificial Bee Colony (ABC) algorithm is inspired by the foraging behavior of honey bees. One example of a naturally occurring swarm that makes a coordinated and intelligent hunt for food is the honeybee swarm. The ability to transmit information, memorize the environment, store and disseminate knowledge, and make decisions based on that knowledge are just a few of the traits of the honeybee swarm. The swarm advances via social learning and teaching and dynamically assigns tasks in response to environmental changes.

Since the ABC algorithm is a population-based optimization technique that assesses fitness, it is expected that the population of potential solutions will slant toward the higher-fitness regions of the search space. Population-based optimization algorithms provide near-optimal solutions to difficult optimization problems through natural motivation. Swarm-based optimization algorithms find solutions through cooperative trial and error methods. Social colonies' peer-to-peer learning behavior drives the ABC optimization algorithms. ABC uses an iterative procedure to identify the best answer from a population of feasible ones [4–7]. The two essential factors that determine the development of an ABC population are variation and selection. The variation process explores diverse sections of the search space. The selection procedure guarantees that experiences are utilized.

The four steps of the ABC algorithm are the scout and onlooker bees phase, the employed bees phase, and the startup phase [8]. ABC creates a uniformly distributed population of solutions, each of which is a dimensional vector, during the population initialization process. The number of variables in the optimization problem for a specific food supply in the population determines how many dimensions there are. Based on the knowledge gained from personal experiences and the new solution's fitness value, the hired bees adjust the existing solution. The bee updates its position with the new food source and discards the old one if the fitness value of the new food source is higher than that of the old one [9–11]. The dimensional vectors that were established previously in the first phase are used to update the position with the number of steps required to obtain the new locations. A random number between − 1 and 1 is the step size.

The process of updating positions during the employed bee phase is illustrated in the adjacent diagram. The search space is two-dimensional. The direction selected at random is shown by the highlighted box, while the present position of a bee is shown by X_i. X_k is the randomly selected bee (Fig. 1). In this phase, the direction of a random bee is deducted from the specified group of bees' direction. The step size, a random integer, exacerbates this discrepancy. Lastly, this number is added to the dimensional vector of X_i to determine the dimension of the new food position 'V'. This vector has the same additional dimensions as X_i and is formed close to X_i. The employed bees share their position and fitness information (nectar) of the updated solutions (food sources) with the hive's spectator bees. Observer bees assess the available information and select a solution based on how likely it is to be fit. Similar to the worker bee, the observer bee updates its memory location and determines whether the potential source is appropriate. If the new location has a greater fitness level than the old one, the bee memorizes it and forgets the old one [12].

If a food source's position is not updated for a predetermined number of cycles, it is considered abandoned. Within the search space, a randomly chosen food source

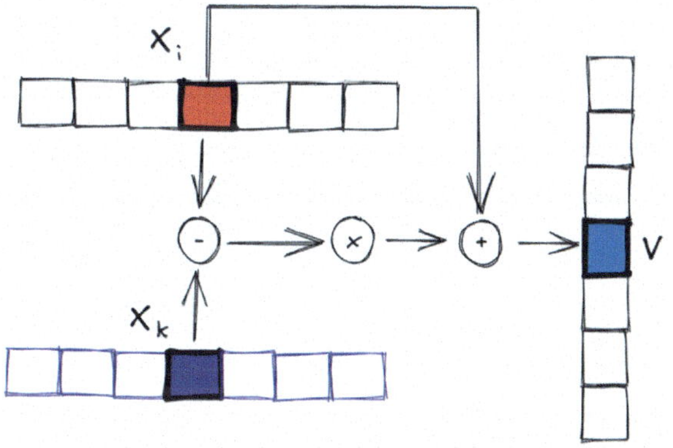

Fig. 1 Procedure for employed bee phase

replaces the abandoned food source, whose bee changes into a scout bee. One important control parameter is the number of cycles or the limit for abandonment in ABC. The working process of the ABC algorithm is shown in Fig. 2.

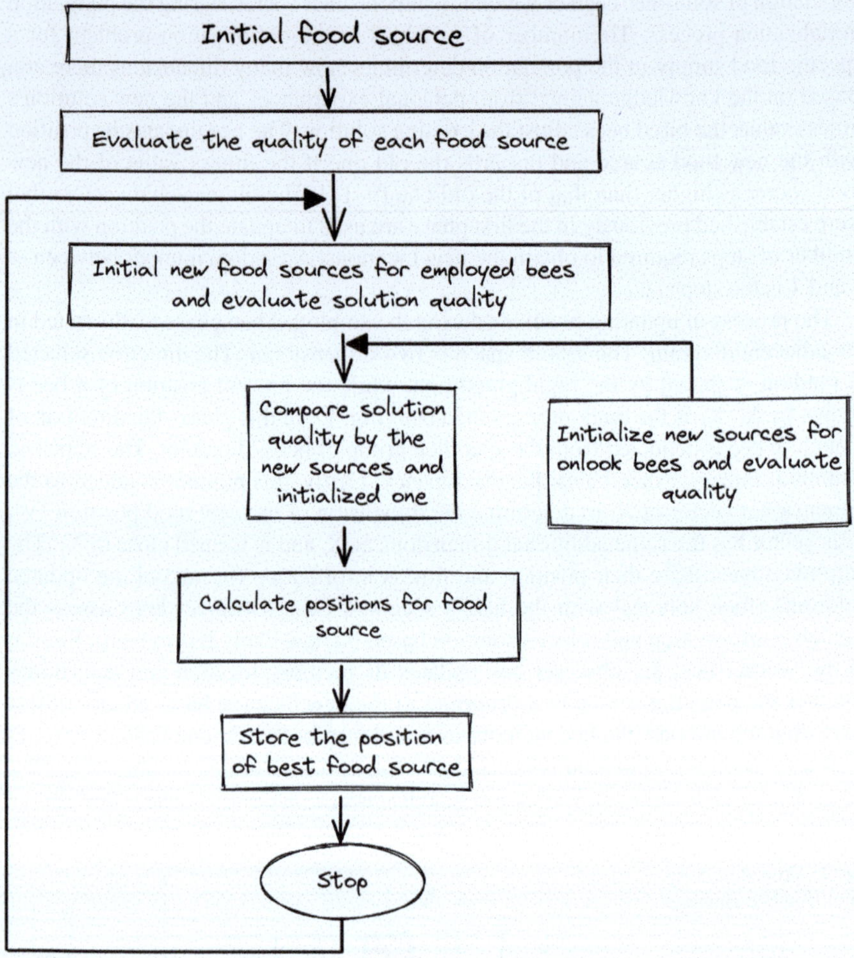

Fig. 2 Working process of ABC

2.1 Inspiration from Bee Behavior and Collective Intelligence

Considering their small brain sizes, bees provide an intriguing case study in collective intelligence that spurs both scientific research and more general considerations about possible social structures.

Efficiency and Resilience: Picture a society in which people work together harmoniously, apply distributed intellect to solve issues, and naturally adjust to change. Bees are a good example of this. They communicate the locations of resources with exceptional accuracy through the waggle dance. Scout bees search for suitable nest locations on their own, and then use their "votes"—dances and vibrations—to draw other bees to the top choice. Since there is no single point of failure in this decentralized decision-making process, efficiency and robustness are guaranteed. Consider using this approach to human systems, where a variety of viewpoints and dispersed information could result in more robust and knowledgeable choices.

Harmony and Sustainability: Bee societies demonstrate an inspiring equilibrium that goes beyond efficiency. Every bee is essential to the community as they don't compete with one another for supremacy. This promotes a peaceful and long-lasting system in which optimal resource management and a link between a person's well-being and the colony's general health exist. Human issues like social cohesion and resource shortage are reflected in this. Could models inspired by bees guide organizational structures that place a high value on cooperation, shared wealth, and long-term sustainability?

Through the study of bee behavior, we can get glimpses of efficient, harmonious communities where the good of the whole is strengthened by the efforts of each individual. Bees are a great source of inspiration for many fields, from social organization theories to problem-solving algorithms, and they can provide invaluable insights for creating a better future.

2.2 Key Characteristics of ABC Algorithms

Adaptation: Memory and Learning: ABC algorithms can adjust their search based on past successes and failures because they keep a "memory" of previously investigated answers. Mechanisms such as employed bees learning the locations of food sources by heart and bystanders selecting sources based on the knowledge of employed bees help to promote this.

Parameter Changes: Throughout the search process, control factors like crossover and mutation rates can be dynamically changed. This aids in the algorithm's ongoing balancing of exploration and exploitation [13–15].

Self-organization: Decentralized Decision-Making: ABC functions without a leader, in contrast to algorithms under central supervision. The working bees, bystanders, and scout bees freely choose which food sources to investigate and use.

Emergent Behavior: Without any conscious coordination, individual bee interactions result in complex collective behaviors like figuring out which food source is preferable.

Exploration–Exploitation Balance: Worker Bees and Observer Bees: Worker bees are explorers; they concentrate on previously identified food sources. In order to encourage the exploitation of promising areas, observer bees balance exploration by selecting sources based on employed bee success.

Scout Bees: Scout bees are a symbol of genuine discovery since they scour uncharted territory for brand-new food sources. By doing this, the algorithm is kept from becoming trapped in local optima.

Exploration is introduced and premature convergence is avoided when randomness is present in ABC by using stochasticity. Parameter tweaking is a key component of ABC's efficacy, parameter-tuning calls for precise modification for given issues. In order to effectively find the best answers to challenging optimization issues, ABC algorithms incorporate traits like adaptation, self-organization, and the exploration–exploitation balance found in bees' foraging behavior. When compared to conventional optimization methods, this bio-inspired strategy has several special advantages [16].

2.3 *Benefits of Using ABC in Diabetic Retinopathy Diagnosis*

The following are some possible advantages of employing ABC algorithms directly for the diagnosis of diabetic retinopathy; however, research in this area is still in its early stages. Enhanced Early Detection and Accuracy is one of its benefits, and ABC algorithms are highly skilled at locating the best answers inside intricate datasets. By using them for retinal image analysis, it may be possible to identify subtle indicators of diabetic retinopathy more accurately, especially in the early stages when treatment has the most benefit. Less Subjectivity and Bias is another benefit of this visual analysis used in traditional diagnosis by ophthalmologists and is susceptible to subjectivity and bias. As impartial instruments for making decisions, ABC algorithms may provide a more standardized and consistent method, lowering false positives and negatives.

Due to simplified workflow and screening, ABC algorithms are fast and effective at analyzing big datasets, and they may be able to automate some of the screening processes. This could increase screening efficiency overall and free up ophthalmologists' time for more complicated patients. Integration with Telemedicine is enabling remote analysis of retinal pictures and possibly increasing access to early diagnosis in disadvantaged areas, and ABC algorithms could be integrated into telemedicine platforms. Potential Personalization has the ability to customize risk assessments and screening frequency for individual patients by integrating patient-specific data such as genetic variables and medical history. However, it is important to consider the limitations such as high-quality data is needed for both training and verifying ABC algorithms, but obtaining sufficient numbers of labeled, high-quality retinal pictures can be costly and difficult. Black box problems can make it difficult to understand how ABC algorithms come to their findings, which could make them less reliable and difficult to comprehend in therapeutic situations. Regulatory permission and thorough testing are necessary for the integration of a new diagnostic tool, and these processes can be difficult and time-consuming. In general, ABC shows promise for enhancing the diagnosis of diabetic retinopathy; nevertheless, more study and development are required to resolve its drawbacks and guarantee its safe and efficient clinical application [17–20].

3 Methodology

3.1 Preprocessing and Analysis of Retinal Images

Retinal images are incredibly useful for detecting and treating a variety of eye illnesses because they depict the complex topography of the back of the eye. However, before a meaningful analysis can be carried out, raw photos frequently need to be preprocessed. Preprocessing sharpens the picture to make it easier to understand, in this diminishing noise improves image clarity by removing pixel blurring, dust artifacts, and inconsistent lighting. Selection of the ROI (region of interest) that focuses on the macula, blood vessels, and optic nerve while ignoring unimportant regions. Contrast enhancement highlights small details that are important for diagnosis by adjusting brightness and color fluctuations [21].

Analysis extracts valuable data from the prepared image that are segmentation, which allows for precise measurements and anomaly detection by isolating particular structures, such as blood vessels or the optic disk. Feature extraction is another one that provides numerical data for study by quantifying pertinent features such as vessel diameter, tortuosity, or the existence of microaneurysms. Classification algorithms can help with diagnosis and decision-making by classifying images as normal or suggestive of specific diseases based on features that have been retrieved [22]. Retinal pictures are snapshots until preprocessing and analysis work together to turn them into potent tools for early disease identification, treatment monitoring, and eventually better patient outcomes.

3.2 Feature Extraction and Selection Using ABC Algorithms

In order to diagnose diabetic retinopathy, feature extraction must detect important visual characteristics of retinal pictures, such as microaneurysms or blood vessel thickness. With hundreds of possible features, it becomes imperative to choose the most informative ones.

ABC algorithms excel in this area, and ABC can use its optimization powers to simplify the feature selection process mainly investigating diverse features that employed bees to forage for good nectar sources by visiting different feature combinations. Scout bees make sure the search doesn't become locked in local optima by investigating completely new features. A classifier (employed bee's "dance") assesses each feature combination's effectiveness in differentiating healthy from diseased eyes [23]. Selective exploitation can favor informational features and gradually hone their search, and onlooker bees select feature combinations based on employed bee success. A subset of features that optimize diagnostic accuracy is produced by this iterative procedure, possibly beating conventional statistical techniques. The potential of ABC to optimize feature selection in complicated datasets such as retinal pictures presents great opportunities for improving the diagnosis of diabetic retinopathy, even though further research is required before real-world implementation can take place [24].

3.3 Datasets Description

There are multiple datasets available for the tasks which consist of retinal images and annotations that are appropriate for preprocessing, analysis, and feature extraction to diagnose diabetic retinopathy and other visual illnesses. Below, we will explore some of the notable datasets in Table 1.

These datasets contain a large number of images and annotations that are essential for creating preprocessing methods, conducting in-depth studies, extracting characteristics, and applying classification techniques. Their role will be crucial in implementing and assessing the effectiveness of the ABC (Artificial Bee Colony) algorithms for extracting and selecting features. Additionally, they will be responsible for incorporating these methods into a diagnostic framework specifically designed for retinal illnesses.

3.4 Integration of ABC with Diagnostic Framework

Imagine a day in the future when the ability to identify diabetic retinopathy is enhanced by intelligence inspired by bees. ABC algorithms integrated with diagnostic frameworks could transform this procedure. Smooth Integration demonstrates

Table 1 Datasets description

Dataset	Description	Contents	Annotations
DIARETDB1	A publicly available database for benchmarking diabetic retinopathy detection algorithms	89 color fundus images, of which 84 contain at least mild non-proliferative signs of diabetic retinopathy, and 5 are normal	Expert annotations of various diabetic retinopathy lesions such as microaneurysms, hemorrhages, hard exudates, and soft exudates
MESSIDOR	A dataset used to develop and evaluate algorithms for automatic detection of diabetic retinopathy and diabetic macular edema	1,200 color fundus images from 654 patients	Grading of diabetic retinopathy and risk of macular edema by medical experts
DRIVE (Digital Retinal Images for Vessel Extraction)	A dataset specifically designed for the development and evaluation of algorithms for retinal vessel segmentation	40 color fundus images (7 abnormal, 33 normal)	Manually labeled segmentations of the blood vessels by two human observers
STARE (Structured Analysis of the Retina)	A dataset for retinal vessel segmentation and the detection of pathologies	400 images, with 20 used for vessel segmentation	Manually segmented vessels and pathology labels
E-Ophtha	A database of fundus images with annotations for exudates and microaneurysms	Two subsets—E-Ophtha EX (47 images with exudates, 35 without) and E-Ophtha MA (148 images with microaneurysms, 233 without)	Expert annotations for the presence of exudates and microaneurysms
Retina Image Bank	A large repository of retinal images for various retinal diseases	Thousands of retinal images depict a wide variety of retinal conditions	Expert annotations for different retinal pathologies
Kaggle Diabetic Retinopathy Detection	A Kaggle competition dataset for diabetic retinopathy detection	35,126 color fundus images	Labels for different stages of diabetic retinopathy

preprocessing and preliminary analysis is performed on retinal pictures. Then, pertinent aspects are taken out and fed into the ABC structure, which serves as the "hive". Collaborative Feature Selection: Employed, onlooker, and scout bees explore various feature combinations, seeking the most informative "nectar" for accurate diagnosis. Their dance language translates to classifier performance, guiding the search toward optimal feature sets [25]. Enhanced Diagnosis being chosen, the features are reintegrated into the diagnostic framework, which may help to improve the performance of current algorithms or fuel the development of brand-new machine learning models for more accurate illness diagnosis [26].

There are various benefits to this integration, and Increased Accuracy is ABC's objective; the feature selection process can perform better than conventional techniques, resulting in diagnoses with fewer false positives and negatives. Interpretability may learn which characteristics most influence the diagnosis by following the "dance" of employed bees, which promotes trust and comprehension in the decision-making process [27]. ABC's adaptability enables it to adjust to fresh datasets and changing illness patterns, guaranteeing the stability and currency of the diagnostic framework. Even if there are still issues with data quality and regulatory barriers, it is clear that this bee-inspired method has the potential to completely change the way that diabetic retinopathy is diagnosed. For innumerable patients, integrating ABC can result in earlier identification, better treatment outcomes, and ultimately, better lives [28].

Numerous studies have tried to use DL to automatically detect and classify DR lesions [29]. We can divide these techniques into four categories: binary, multi-level, lesion-based, and vessel-based, depending on the type of classification approach employed (Table 2).

4 Conclusion

Finally, the integration of Artificial Bee Colony (ABC) algorithms with medical image analysis has made a significant advance in the identification of diabetic retinopathy. This methodology improves diagnosis precision, reduces computational complexity, and facilitates instantaneous analysis, resulting in improved patient outcomes and increased healthcare efficiency. The system's ability to adapt to various imaging techniques and datasets highlights its potential for extensive use in clinical settings. This innovative approach uses the adaptable and self-organizing features of ABC algorithms to improve optical diagnostics, resulting in improved patient treatment and a decrease in vision impairment caused by diabetic retinopathy.

Table 2 Performance of various deep learning methods

S. No	Deep learning methods	Lesion detection	Dataset size	Performance measure			
				AUC	Accuracy	Sensitivity	Specificity
1	CNN (Inception-v3)	No	Messidor-2 (1748) and EyePACS-1 (9963)	–	–	96.7% 98.5%	94.9% 93.9%
2	CNN	Yes	Messidor-2 (1748)	0.980	–	95.9%	88.0%
3	CNN	No	Kaggle (80,000)	–	76%	40%	96%
4	CNN	No	Kaggle (1000)	–	94.7%	–	–
5	CNN	Yes	Kaggle (88,702), DiaretDB1 (89), and E-ophtha (107,799)	0.966–0.958	–	–	–
6	CNN	Red lesion only	DIARETDB1 (89), E-Ophtha (381), and MESSIDOR (1200)	CPM = 0.5487 for DIARETDB1 and CPM = 0.4357 for e-ophtha		0.4883 0.3680	–
7	CNN-ResNet34	No	Kaggle (35,000)	–	87%	88%	–
8	DNN, CNN (VGGNET architecture), BNN	No	Kaggle (2000)	–	BNN = 44% DNN = 87.3% CNN = 79.3%	–	–
9	CNN (InceptionNet V3, AlexNet, and VGG16)	No	Kaggle (166)	–	AlexNet = 38.84%, VGG16 = 52.03%, and InceptionNet V3 = 64.23%	–	–
10	CNN (AlexNet, VggNet, GoogleNet, and ResNet)	No	Kaggle (35,126)	The higher is VggNet-s (0.9786)	The higher is VggNet-s (96.68%)	VggNet-16 achieved higher result (91.78%)	The higher is VggNet-s (98.43%)
11	CNN	MA only	E-Ophtha (381), DRIVE (100), and DIARETDB1 (89)	0. 625 0.219 0.489	–	–	–

(continued)

Table 2 (continued)

S. No	Deep learning methods	Lesion detection	Dataset size	Performance measure			
				AUC	Accuracy	Sensitivity	Specificity
12	CNN	EX only	DiaretDB0 (130), DiaretDB1 (89), and Diabetic Retinopathy Detection (125)	–	99.17 98.53 99.18	99.98 99.2 99.9	98.76 97.89 98.54
13	Fully CNN	No	STARE (20), HRF (45), DRIVE (40), and Retina Image Bank (28)	0.9905 0.9801 0.9823 0.9875	0.9734 0.9758 0.9678 0.9687	0.8590 0.7876 0.7989 0.7674	0.9887 0.9756 0.9867 0.9856
14	Fully CNN	No	STARE (20), DRIVE (40), and Diabetic Retinopathy Detection (28)	0.9905 0.9845 0.9878	0.9694 0.9576 0.9723	0.8435 0.8245 0.7779	0.9858 0.9804 0.9864

References

1. Qawqzeh Y, Alharbi MT, Jaradat A, Abdul Sattar KN (2021) A review of swarm intelligence algorithms deployment for scheduling and optimization in cloud computing environments. PeerJ Comput Sci 7. https://doi.org/10.7717/peerj-cs.696
2. Arcadu F, Benmansour F, Maunz A et al (2019) Deep learning algorithm predicts diabetic retinopathy progression in individual patients. npj Digit Med 2:92. https://doi.org/10.1038/s41746-019-0172-3
3. Raza K, Singh NK (2021) A tour of unsupervised deep learning for medical image analysis. Curr Med Imag 17(9):1059–1077
4. Alsaidy SA, Abbood AD, Sahib MA (2020) Heuristic initialization of PSO task scheduling algorithm in cloud computing. J King Saud Univ. https://doi.org/10.1016/j.jksuci.2020.11.002. Epub ahead of print Nov 13 2020
5. Aslan S (2019) A transition control mechanism for artificial bee colony (ABC) algorithm. Comput Intell Neurosci 5012313. https://doi.org/10.1155/2019/5012313
6. Aslan S, Karaboga D (2020) A genetic Artificial Bee Colony algorithm for signal reconstruction based big data optimization. Appl Soft Comput J 88:106053–106053
7. Beed R, Roy A, Bhattacharya D (2019) A hybrid multi-objective tour route optimization algorithm based on particle swarm optimization and artificial bee colony optimization. Comput Intell 88:884–909
8. Dai W, Liang L, Zhang B (2020) Firefly optimization algorithm for the prediction of uplift due to high-pressure jet grouting. Adv Civ Eng:8833784. https://doi.org/10.1155/2020/8833784
9. Dai X, Long S, Zhang Z, Gong D (2019) Mobile robot path planning based on ant colony algorithm with a * heuristic method. Front Neurorobot 13:15. https://doi.org/10.3389/fnbot.2019.00015
10. Dokeroglu T, Sevinc E, Cosar A (2019) Artificial bee colony optimization for the quadratic assignment problem. Appl Soft Comput J 76:595–606. https://doi.org/10.1016/j.asoc.2019.01.001
11. Wani N, Raza K (2018) Chapter 3—Multiple Kernel-learning approach for medical image analysis. In: Dey N, Ashour AS, Shi F, Balas VE (eds) Soft computing based medical image analysis. Academic Press, pp 31–47. ISBN 9780128130872. https://doi.org/10.1016/B978-0-12-813087-2.00002-6
12. Ewees AA, Al-qaness M, Elaziz MA (2021) Enhanced salp swarm algorithm based on firefly algorithm for unrelated parallel machine scheduling with setup times. Appl Math Modell 94:285–305. https://doi.org/10.1016/j.apm.2021.01.017
13. Farahlina Johari N, Mohd Zain A, Haszlinna Mustaffa N, Udin A (2017) Machining parameters optimization using hybrid firefly algorithm and particle swarm optimization. J Phys. Conf Series 892(1):012005. https://doi.org/10.1088/1742-6596/892/1/012005
14. Farid M, Latip R, Hussin M, Hamid N (2020) A survey on QoS requirements based on particle swarm optimization scheduling techniques for workflow scheduling in cloud computing. Symmetry (Basel) 12(4). https://doi.org/10.3390/SYM12040551
15. Hussein BM, Jaber AS (2020) Unit commitment based on modified firefly algorithm. Meas Control (United Kingdom) 53(3–4):320–327. https://doi.org/10.1177/0020294019890630
16. Raza K, Singh S (2024) Artificial intelligence and autoimmune diseases. In: Studies in computational intelligence, vol 1133. Springer
17. Khan A, Hizam H, Wahab N, Othman ML (2020) Optimal power flow using hybrid firefly and particle swarm optimization algorithm. PLOS ONE 15(8):1–21. https://doi.org/10.1371/journal.pone.0235668
18. Kruekaew B, Kimpan W (2020) Enhancing of artificial bee colony algorithm for virtual machine scheduling and load balancing problem in cloud computing. Int J Comput Intell Syst 13(1):496–510. https://doi.org/10.2991/ijcis.d.200410.002
19. Li X, Gao R, Yu L, Li N, Liu J, Wang Y, Feng E (2021) Memetic strategy of particle swarm optimization for one-dimensional magnetotelluric inversions. Mathematics 9(519):1–22

20. Nayak J, Naik B, Dinesh P, Vakula K, Dash PB (2020) Firefly algorithm in biomedical and health care: advances, issues and challenges. SN Comput Sci 1(6):311. https://doi.org/10.1007/s42979-020-00320-x
21. Pradhan A, Bisoy SK (2020) A novel load balancing technique for cloud computing platform based on PSO. J King Saud Univ. https://doi.org/10.1016/j.jksuci.2020.10.016. Epub ahead of print Oct 22 2020
22. Shahabi Sani N, Manthouri M, Farivar F (2020) A multi-objective ant colony optimization algorithm for community detection in complex networks. J Ambient Intell Human Comput 11(1):5–21. https://doi.org/10.1007/s12652-018-1159-7
23. Sheoran S, Mittal N, Gelbukh A (2020) Artificial Bee Colony algorithm in data flow testing for optimal test suite generation. Int J Syst Assurance Eng Manag 11(2):340–349. https://doi.org/10.1007/s13198-019-00862-1
24. Tabrizchi H, Kuchaki Rafsanjani M, Balas VE (2021) Multi-task scheduling algorithm based on self-adaptive hybrid ICA–PSO algorithm in cloud environment. Adv Intell Syst Comput 1222:422–431. https://doi.org/10.1007/978-3-030-52190-5_30
25. Tamura Y, Sakiyama T, Arizono I (2021) Ant colony optimization using common social information and self-memory. Complexity. https://doi.org/10.1155/2021/6610670
26. Windarto W, Eridani E (2020) Comparison of particle swarm optimization and firefly algorithm in parameter estimation of lotka-volterra. AIP Conference Proceedings 2268(1):050008. https://doi.org/10.1063/5.0017245
27. Yu H (2020) Evaluation of cloud computing resource scheduling based on improved optimization algorithm. Complex Intell Syst 0123456789:1–6. https://doi.org/10.1007/s40747-020-00163-2
28. Zhang G, Wang H, Zhao W, Guan Z, Li P (2021) Application of improved multi-objective ant colony optimization algorithm in ship weather routing. J Ocean Univ China 20:45–55. https://doi.org/10.1007/s11802-021-4436-6
29. Zhao R, Liu Q, Li C, Wang Y, Dong D (2020) Performance comparison and application of SI algorithms in crowd evacuation. In: ACM international conference proceeding series, pp 47–51

Case Studies

Artificial Bee Colony Algorithms in Gene Expression Studies: A Case Study

Shaban Ahmad, Khushnuma Tanveer, Nagmi Bano, Faiyaz Ahmad, and Khalid Raza

1 Introduction

In bioinformatics, deciphering genetic information is a cornerstone in understanding the intricate mechanisms managing life processes [17]. The advent of high-throughput technologies, such as microarrays and next-generation sequencing, has revolutionised the field, enabling the exploration of an organism's genome at an unprecedented scale [41]. This data deluge has presented opportunities and challenges, as genetic information's sheer volume and complexity necessitate innovative computational approaches for its analysis. Within this context, optimisation algorithms have garnered considerable attention for their ability to extract meaningful insights from vast datasets [32]. These algorithms, inspired by natural phenomena or mathematical principles, offer powerful tools for solving complex optimisation problems encountered in various scientific domains. One such algorithm, the artificial bee colony (ABC) algorithm, has demonstrated remarkable efficacy in addressing optimisation challenges, including those inherent in gene expression studies [37].

Gene expression studies entail analysing the transcriptional activity of genes within an organism, providing invaluable insights into cellular processes, disease mechanisms, and developmental pathways. At the core of these studies lies the quantification of mRNA transcripts, which serve as proxies for gene activity levels [19]. High-throughput techniques facilitate the simultaneous measurement of gene expression levels across thousands of genes, yielding comprehensive snapshots of cellular states under different conditions. The microarray technology, for instance, allows the probe gene expression patterns by hybridising labelled RNA samples to complementary probes immobilised on a solid substrate. Subsequent detection and quantification of bound transcripts reveal differential expression profiles indicative of cellular responses to stimuli or pathological conditions [52]. Similarly,

S. Ahmad · K. Tanveer · N. Bano · F. Ahmad · K. Raza (✉)
Department of Computer Science, Jamia Millia Islamia, New Delhi, India
e-mail: kraza@jmi.ac.in

K. Raza (ed.), *Solving with Bees*, Springer Tracts in Nature-Inspired Computing,
https://doi.org/10.1007/978-981-97-7344-2_10

next-generation sequencing methodologies offer nucleotide-level resolution of transcriptomes, enabling precise quantification and characterisation of RNA species. The analysis of gene expression data entails several vital steps, including preprocessing, normalisation, differential expression analysis, and functional interpretation. Preprocessing involves quality assessment, background correction, and normalisation procedures to mitigate technical artefacts and ensure data integrity. Following preprocessing, statistical methods are employed to identify genes that exhibit significant changes in expression between experimental conditions, shedding light on biological processes underlying phenotypic differences [33]. Functional interpretation of gene expression data relies on enrichment analysis techniques, which influence curated databases to identify overrepresented biological pathways, molecular functions, or cellular components associated with differentially expressed genes. Such analyses facilitate the elucidation of biological themes and the prioritisation of candidate genes for further investigation [33].

The analysis of gene expression data often entails optimising complex computational tasks, including feature selection, classification, clustering, and pathway analysis [43]. These tasks pose formidable challenges due to the high-dimensional nature of gene expression datasets, characterised by many features (genes) relative to the number of samples [39]. Optimisation algorithms play a pivotal role in addressing these challenges by efficiently exploring solution spaces to identify optimal or near-optimal solutions [57]. These algorithms encompass various methodologies, from evolutionary and swarm intelligence-based approaches to gradient-based optimisation techniques. Among these algorithms, the ABC algorithm has emerged as a promising tool for addressing optimisation problems in gene expression studies. Inspired by the foraging behaviour of honeybee colonies, ABC algorithms harness the collective intelligence of artificial bees to search for optimal solutions in complex problem domains [48]. The fundamental premise of ABC algorithms lies in emulating the foraging behaviour exhibited by honeybees in nature. In a typical ABC algorithm, a population of artificial bees traverses a multidimensional search space, iteratively exploring potential solutions while exchanging information through a process analogous to the waggle dance employed by natural bees to communicate the location of food sources. At each iteration, artificial bees (employed bees, onlooker bees, and scout bees) explore candidate solutions within their local neighbourhoods, guided by the quality of solutions and the information gleaned from their interactions with other bees [4]. Through this process of exploration and exploitation, ABC algorithms iteratively refine candidate solutions, converging towards optimal or near-optimal solutions to the underlying optimisation problem. The efficacy of ABC algorithms stems from their ability to balance exploration and exploitation, effectively navigating complex solution spaces characterised by nonlinearity, multimodality, and high dimensionality. By leveraging principles of swarm intelligence and stochastic search, ABC algorithms offer robust and scalable solutions to a wide range of optimisation problems, including those encountered in gene expression analysis. In the context of gene expression studies, ABC algorithms find applications in diverse areas such as feature selection, model parameter tuning, clustering analysis, and pathway enrichment. By efficiently navigating the vast search spaces inherent in these tasks, ABC algorithms enable the identification of informative gene subsets, optimising classification models, and elucidating underlying biological mechanisms driving differential gene expression patterns [6].

Furthermore, ABC algorithms exhibit several desirable characteristics that make them well-suited for optimisation tasks in gene expression analysis. These include their ability to handle high-dimensional and noisy data, their robustness to local optima, and their capacity for parallelisation and scalability. As such, ABC algorithms represent valuable tools to extract meaningful insights from gene expression data and unravel the complexities of biological systems. Optimisation algorithms, particularly ABC algorithms, are crucial in advancing our understanding of gene expression patterns and their implications for biological processes and disease mechanisms [2, 27, 56, 58]. By providing efficient and scalable solutions to optimisation problems in gene expression analysis, these algorithms empower the extraction of actionable knowledge from vast and complex datasets, ultimately driving innovation in biomedical research and personalised medicine. Nature-inspired optimisation algorithms are classified as Bio-inspired, Physics and Chemistry-based algorithms, and others (Fig. 1).

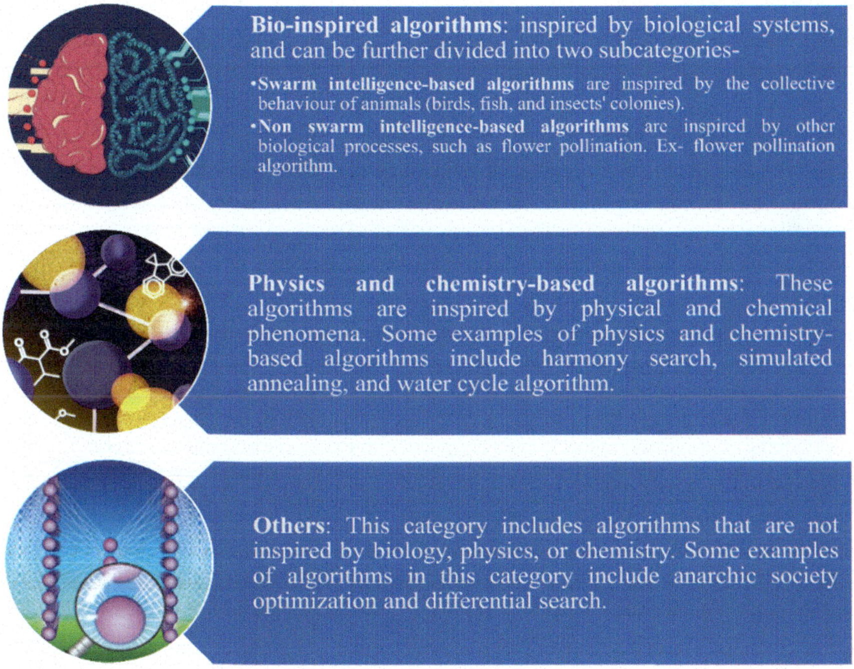

Fig. 1 Representation of various Nature-inspired optimisation algorithms

2 Literature Review

In this era of burgeoning biological data and complex optimisation problems, applying optimisation techniques in bioinformatics, particularly in gene expression studies, holds immense promise for advancing our understanding of biological systems and disease mechanisms. Metaheuristic optimisation algorithms have emerged as powerful tools for tackling global optimisation problems across various research domains, including data science, systems biology, and cancer research. Among the plethora of optimisation techniques available, stochastic algorithms like Particle Swarm Optimisation (PSO), Genetic Algorithm (GA), Ant Colony Optimisation (ACO), Harmony Search (HS), biogeography-based optimisation (BBO), and ABC algorithms have demonstrated exceptional efficacy in addressing a wide range of optimisation challenges [11, 12]. These algorithms have exhibited superior performance in combinatorial optimisation, neural network training, image processing, pattern recognition, data mining, molecular docking, and other domains, making them invaluable tools [13]. In gene expression analysis, optimisation problems abound, from gene selection to high-dimensional microarray data allocation. Metaheuristic algorithms, including ABC algorithms, are extensively utilised to solve these combinatorial optimisation problems, enabling advancements in medical image analysis, gene selection for classification, cancer diagnosis, and prognosis. By harnessing the collective intelligence of artificial bees and other metaheuristic algorithms, we can navigate the complex landscape of genomic data, identify relevant biomarkers, and uncover novel insights into disease mechanisms. As we move forward, integrating optimisation techniques with gene expression studies will continue to drive innovation and revolutionise biomedical research [8]. By addressing the challenges and embracing future research directions outlined in this case study, we can unlock the full potential of optimisation algorithms in deciphering the intricacies of gene expression regulation and translating this knowledge into improved diagnostics, therapeutics, and personalised medicine. Optimisation algorithms and gene expression studies pave the way for a deeper understanding of developing innovative solutions to address pressing healthcare challenges [33].

2.1 Existing Approaches in Gene Expression Studies

Gene expression analysis has received attention in bioinformatics, cancer informatics, and biotechnology, and its performance has not been addressed in several high-dimensional datasets. Data on gene expression is stored in the microarray data, which classifies and diagnoses genetic diseases. The prediction of significant genes is essential for cancer studies, and bio-inspired algorithms have been widely used [16, 35].

2.1.1 Genetic Bee Colony

It is a hybrid approach based on the ABC and GA naturally inspired algorithm to select the most informative and predictive genes from a microarray gene-expression profile for cancer classification [5]. Many studies have presented plenty of feature selection strategies, which are analysed and examined over high-dimensional datasets, and several metaheuristic algorithms have been discovered for gene expression case studies, such as Genetic Bee Colony (GBC), Cuckoo search algorithm, ABC, and Binary particle swarm optimisation [50, 46, 54, 58]. GBC is comprised of five phases, which are in Fig. 2.

The ABC parameter settings should be initialised, including the main parameters, population size (PS), number of bees (size = 2 PS), and L is the parameter limit. Further, the solution size is equal to PS initialised randomly by the given equation [15].

$$u = u_{i,j}^{\min} + \text{rand}[0,1]\left(u_{i,j}^{\max} - u_{i,j}^{\min}\right)$$

where i represents the solution index, j represents the decision variable, and the random value equals 0 and 1. The obtained generated solution is determined using objective functions [15]. The neighbourhood solution can be extracted by using the following equation:

$$v_{i,j} = u_{i,j} + \theta_{i,j}\left(u_{i,j} - u_{k,j}\right)$$

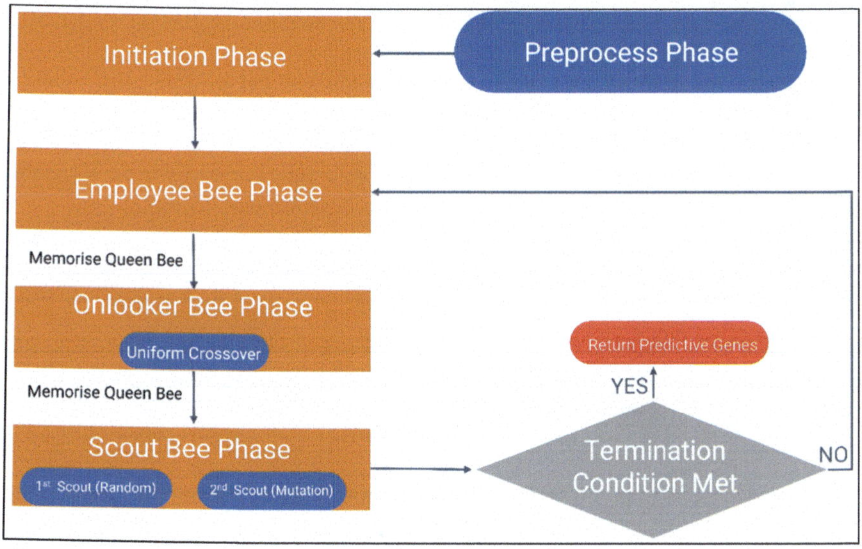

Fig. 2 The crucial phases of the Genetic Bee Colony Algorithm

The exploitation of fitness value can be done through Onlooker bee according to the following equation:

$$p_i = \frac{f_i t_i}{\sum_{j=1}^{PS} f_i t_j}$$

Then, ABC algorithm equations i and ii are modified to generate an initial solution.

$$U_i : i = 1, \ldots, \mathrm{SN} u_{i,j} = \begin{cases} 0, \text{if} G(0{,}1) \leq 0.5 \\ 1, \text{if} G(0{,}1) > 0.5 \end{cases}$$

2.1.2 Binary Particle Swarm Optimisation Algorithm

In 1997, Kenned and Eberhart proposed a stochastic population-based optimisation algorithm known as Particle Swarm optimisation. Implementing the Binary Particle Swarm algorithm, BPSO is utilised to extract modified genes compared to patients who have gone under investigation. BPSO is based on three following steps:

- Firstly, velocity and the position of particles are generated.
- The velocity is upgraded.
- Upgrading particle positions.
- The velocity of every particle can be upgraded as follows:

$$V_{\mathrm{id}}^{\mathrm{new}} = w \cdot V_{\mathrm{id}}^{\mathrm{old}} + c_1 \mathrm{rand}_1(\ldots) x(\mathrm{pb}_{\mathrm{id}}^{\mathrm{old}} - x_{\mathrm{id}}^{\mathrm{old}}) + c_2 \mathrm{rand}_2(\ldots) x(\mathrm{gb}_{\mathrm{id}}^{\mathrm{old}} - x_{\mathrm{id}}^{\mathrm{old}})$$

c_1-Cognitive learning factor (0–4), c_2-Social Learning Factor (0–4), the value of inertia weight w is used to decrease the swarm control and velocity of the particle ($w = 0.4$–0.9), and rand_1 and rand_2 are distributed between 0 and 1 [55]. To avoid particles' sharp movement in the search space, velocities are tied within V_{min} and V_{max}. The particular position is updated by using the following formula:

$$x_{\mathrm{id}}^{\mathrm{new}} = x_{\mathrm{id}}^{\mathrm{old}} + x_{\mathrm{id}}^{\mathrm{new}}$$

The following equation updates the dimension:

$$x_i^{\mathrm{new}} = \begin{cases} 1, \text{ if sigmoid}\left(v_i^{\mathrm{new}}\right) > \mathrm{rand}(0{,}1) \\ \quad 0, \text{ otherwise} \end{cases}$$

The procedure is implemented in tenfold rotation estimation to contribute to the entire training set in the abstraction of the most discriminative genes [22]. The insertion of every gene in the finest set depends on the quotability of that gene from the entire number of folds. The maximum quotable number had been set to 10 so that some highly accurate genes could be inserted into the set of genes.

2.1.3 Genetic Algorithm

The genetic algorithm GA, inspired by natural selection, is a metaheuristic. An evolutionary algorithm is used to analyse microarray data, which stores information on gene expression [11]. In this technique, mutation, crossover, and selection operators are used to search for the finest results from individuals. The initial steps are to select chromosomes with high adapting values through the selection operator, followed by the crossover operator, which combines two chromosomes from which the offspring is generated. Finally, the mutation operator modifies the estimated value of more chromosome genes from their starting stages. This procedure is repeated to acquire the finest fitness value and determine the quality of every chromosome. The classification precision of the classifier estimates the fitness of all gene subsets. The gene subset that has the best 10 cross-validation classification precision is the deliberated optimal gene subset [24]. In Fig. 3, we show the flowchart of genetic algorithms.

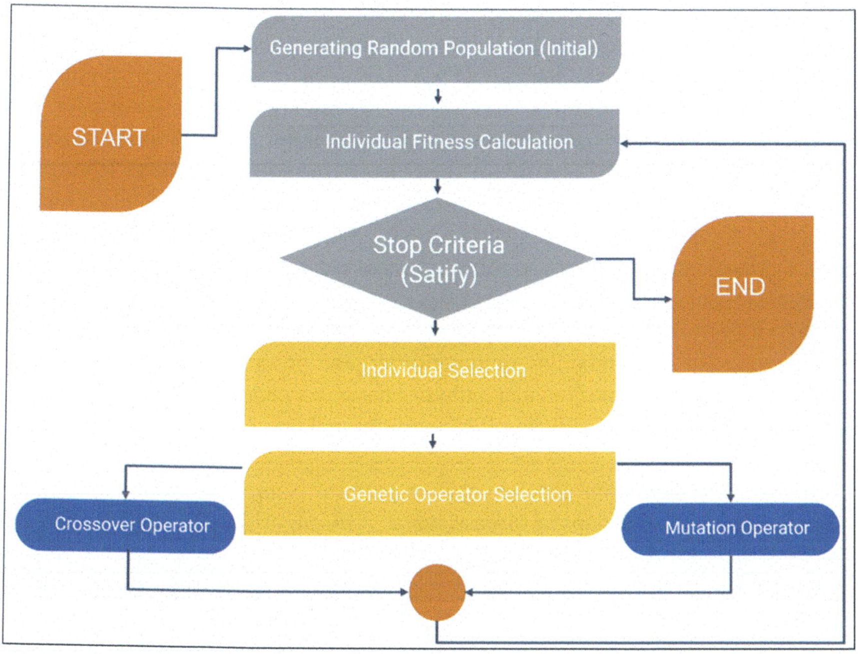

Fig. 3 Showing the flowchart of Genetic Algorithm

2.2 *Status of ABC in Gene Expression Analysis*

The feeding habits of honeybee swarms serve as the model for population-based metaheuristic optimisation approaches known as ABC algorithms [26]. By simulating the search procedures of actual bees, such as worker, observer, and scout bees, a population of artificial bees in ABC algorithms iteratively searches the solution space to identify optimum solutions. ABC algorithms' simplicity, effectiveness, and capacity to manage challenging optimisation issues, including nonlinear and multimodal search spaces, define them [21].

Gene expression analysis is crucial in understanding the mechanisms behind biological processes, such as disease development, medication discovery, and customised therapy. Large-scale data on gene expression is being produced by high-throughput technologies like next-generation sequencing and microarrays, which calls for sophisticated computational methods to analyse the data [42]. Gene expression data analysis presents several obstacles, and ABC methods have shown promise as an optimisation method. Numerous investigations have examined the effectiveness of ABC algorithms in various elements of gene expression analysis, providing insight into their advantages, disadvantages, and possible areas for advancement. Numerous applications of ABC algorithms in gene expression investigations are covered by the literature study, including but not limited to the following:

Feature Selection: ABC algorithms have been used to pick features from gene expression data to find pertinent genes linked to particular biological situations or phenotypes. Research findings indicate that ABC-based feature selection techniques enhance classification precision, diminish dimensionality, and pinpoint biomarkers for diagnosing and prognosis illnesses [3].

Gene Clustering: Gene expression data was clustered using ABC methods to find co-expressed gene modules or functional gene groupings. ABC-based clustering algorithms can identify gene regulatory networks and functional pathways by dividing genes into biologically significant groups and optimising clustering criteria [31].

Classification: Based on gene expression patterns, samples will be categorised into distinct biological groups or disease states. ABC algorithms have been used for these classification tasks in gene expression analysis. Compared to conventional machine learning methods, ABC-based classification models have demonstrated competitive performance, especially when a lack of training data or gene expression measurements is noisy [9].

Regulatory Network Inference: To decipher the intricate relationships between genes and their regulatory components, ABC algorithms have been used to infer gene regulatory networks from gene expression data. ABC methods can identify putative regulatory linkages, transcription factor binding sites, and signalling pathways underpinning gene expression dynamics by optimising network inference techniques [9].

2.3 Limitations and Challenges

The requirement for meticulous parameter adjustment is one of the main difficulties encountered using ABC algorithms in gene expression research. The right parameter settings, such as colony size, scout bee count, and rates of exploration and exploitation, are critical to the effectiveness of ABC algorithms [49]. Choosing the best parameter values can be difficult because it frequently necessitates a thorough understanding of both the method and the issue area. Furthermore, ABC algorithms' sensitivity to parameter settings might lead to premature convergence or inadequate search space exploration, impairing their ability to effectively identify the underlying patterns in gene expression data [31].

The intrinsic stochastic character of ABC algorithms is another drawback. There is no certainty of discovering the globally optimal solution since ABC algorithms involve stochastic processes, such as random initialisation of solutions and probabilistic decision-making by scout bees during the search process [54]. The algorithm's stochastic tendency can produce inconsistent results after several runs, making it difficult to evaluate the stability and dependability of the solutions produced. Moreover, reproducibility—a crucial prerequisite for scientific research, especially in gene expression analysis, where the validity of findings depends on the consistency of results across various experiments—can be compromised by the stochastic nature of ABC algorithms. Furthermore, working with large-scale gene expression datasets is a considerable problem due to the scalability of ABC techniques. Applying conventional ABC techniques directly becomes unfeasible as the dataset size increases due to the exponential growth in computational complexity of the optimisation issue. Large datasets can be handled by scaling up ABC algorithms using creative parallelisation strategies, distributed computing, or hybrid systems combining ABC with other optimisation techniques to improve scalability without sacrificing solution quality [31].

3 ABC Algorithms

ABC algorithms have emerged as a powerful tool in optimisation problems, drawing inspiration from the foraging behaviour of honeybee colonies [12]. We provide an overview of ABC algorithms and then explore their working mechanism. Furthermore, we discuss the advantages and limitations of ABC algorithms in optimisation, shedding light on their potential and areas for improvement.

3.1 *Overview of ABC Algorithms*

The population-based optimisation method known as ABC algorithms was developed in response to the foraging habits of honeybee colonies. Since their introduction by Karaboga in 2005, ABC algorithms have become more well-liked because of their adaptability, efficiency, and capacity to tackle challenging optimisation issues [8]. ABC algorithms are based on modelling honeybee foraging behaviour, which includes communication, exploitation, and exploration within a colony. Three sorts of artificial bees comprise the population at the centre of ABC algorithms: scout, employed, and observer bees. While spectator bees choose viable solutions based on the information supplied by employed bees, employed bees investigate prospective solutions inside the search space. On the other hand, scout bees are assigned to investigate uncharted territory in the search space if employed bees cannot yield satisfactory results [31].

ABC algorithms use a trial-and-error methodology in which artificial bees exchange information to develop solutions iteratively. Every iteration or cycle has three primary phases: the employed bee period, the spectator bee phase, and the scout bee phase. Employed bees investigate the area surrounding their present solutions and assess their suitability during the employed bee phase. After that, promising solutions are disseminated among observing bees, who use fitness values to make a probabilistic selection of the answers. Scout bees regularly investigate novel solutions by producing potential solutions randomly throughout the search space [29].

3.2 *Working Mechanism of ABC Algorithms*

The interaction and cooperation of artificial bees to find the best solutions within a given optimisation problem is the basis for the operation of ABC algorithms [26]. The first step in the procedure is to initialise a population of artificial bees, each of which stands for a possible answer somewhere in the search area. The solutions generated randomly are assigned to employed bees, and their fitness is assessed afterwards. Every employed bee perturbs its position inside the search space to investigate the vicinity of its assigned solution during the employed bee phase. Local search operators like mutation and crossover, which seek to produce new candidate solutions, are used to achieve this disruption. The employed bees assess the suitability of these potential solutions and adjust their locations accordingly, keeping the best ones found thus far. Fitness ratings are used as a guide, and onlooker bees use probability to choose solutions during the onlooker bee phase. Larger fitness solutions are more likely to be chosen, simulating the recruitment pattern in honeybee colonies [36]. Observer bees subsequently assess the chosen solutions and adjust their positions correspondingly, promoting information sharing among artificial bees. Scout bees are essential to the population's ability to remain diverse because they constantly look for new ways to solve problems. Scout bees create fresh candidate solutions randomly within the search space when hired bees cannot deliver suitable solutions or become stationary. Premature convergence is avoided by this exploration technique, which also makes sure that the search space is fully investigated.

3.3 Advantages and Limitations of ABC in Optimisation

ABC algorithms are excellent for handling various optimisation issues because of their many benefits. First, practitioners with different experience levels can use ABC algorithms because they are easy to implement and require little parameter adjusting [14]. Furthermore, since they are reliable and effective, ABC algorithms can handle complicated optimisation problems with high-dimensional search spaces. In addition, compared to conventional optimisation methods, ABC algorithms show good convergence qualities and are less prone to being stuck in local optima. This is explained by the fact that scout bees use various exploration strategies to ensure the search space is thoroughly investigated. Furthermore, ABC methods may scale to more prominent problem instances and efficiently use computer resources since they are naturally parallelisable.

Even with all of their benefits, ABC algorithms have some drawbacks that should be considered. The main drawback of ABC algorithms is their dependence on randomisation, which can result in unpredictable behaviour and performance variability. This unpredictability could make it difficult to assess the algorithm's convergence behaviour and impair the reproducibility of outcomes. Additionally, non-smooth or discontinuous fitness landscape optimisation problems may be complex for ABC algorithms to solve; in these cases, gradient-based optimisation methods work better. Furthermore, the choice of parameter variables, such as the population size and exploration rate, significantly impacts how well ABC algorithms perform. Inadequate parameter settings can lead to unsatisfactory solutions that either prematurely or poorly converge [14, 31].

4 Application of ABC in Gene Expression Studies

Applying ABC algorithms in gene expression studies offers a powerful approach to address optimisation challenges in analysing high-dimensional genomic data. ABC algorithms excel in balancing exploration and exploitation, efficiently navigating complex solution spaces to identify informative gene subsets, optimise classification models, and elucidate underlying biological mechanisms [23]. Their robustness to noisy data, adaptability to high dimensionality, and ease of implementation make them particularly well-suited for gene expression analysis. ABC algorithms have been successfully applied in diverse areas such as feature selection, classification, clustering, and pathway analysis. They have facilitated the identification of biomarkers associated with specific disease states, improved the accuracy of predictive models for clinical outcomes, and enhanced our understanding of molecular pathways underlying complex phenotypes [40]. The application of ABC algorithms in gene expression studies represents a valuable tool for extracting meaningful insights from genomic data and advancing our understanding of biological processes and disease mechanisms.

4.1 Rationale for Choosing ABC in Gene Expression:

The selection of the ABC algorithm for gene expression studies stems from its unique ability to address the optimisation challenges inherent in analysing high-dimensional and noisy datasets. The unique characteristics of ABC algorithms make them a compelling choice for addressing the optimisation challenges encountered in gene expression studies, enabling the extraction of meaningful insights from complex genomic datasets. Several factors contribute to the rationale for choosing ABC in this context:

4.1.1 Exploration–Exploitation Balance

ABC algorithms excel at balancing exploration (broadly searching the solution space) and exploitation (narrowing down to promising regions).

4.1.2 Robustness to Noisy Data

Gene expression datasets often contain noise due to experimental variability or technical artefacts. ABC algorithms are robust to such noise, employing stochastic search mechanisms that effectively filter out irrelevant information and focus on informative patterns [35, 37].

4.1.3 Adaptability to High Dimensionality

Gene expression datasets typically exhibit high dimensionality, with thousands of genes representing potential features. ABC algorithms are well-suited to handle such high-dimensional spaces, as they rely on population-based search strategies that can efficiently explore complex solution landscapes [35, 37].

4.1.4 Ease of Implementation and Parameter Tuning

ABC algorithms are relatively simple to implement and require minimal parameter tuning compared to other optimisation techniques. This simplicity makes them accessible with varying computational expertise and facilitates their integration into existing analytical pipelines for gene expression analysis [35, 37].

4.2 Comparative Analysis with Other Optimisation Techniques

While ABC algorithms offer several advantages for gene expression analysis, comparing their performance with other optimisation techniques commonly employed in this domain is essential. Comparative analyses provide valuable insights into the strengths and limitations of different approaches, aiding in selecting the most appropriate methodology for their specific research questions [45]. One standard benchmark for comparing optimisation techniques in gene expression studies is their performance in feature selection and classification tasks. Feature selection aims to identify subsets of genes most informative for discriminating between different biological conditions, while classification involves building predictive models based on these selected features. In comparative analyses, ABC algorithms have demonstrated competitive performance relative to other optimisation techniques, such as genetic algorithms, particle swarm optimisation, and differential evolution [30]. Studies have shown that ABC algorithms often exhibit comparable or superior performance in terms of solution quality, convergence speed, and robustness to noise and dimensionality. Moreover, ABC algorithms offer unique advantages over traditional optimisation techniques, such as their ability to handle multimodality, implementation simplicity, and parallelisation capacity. These features make ABC algorithms particularly well-suited for optimisation tasks in gene expression analysis, where the search space is complex and high-dimensional. However, it is essential to note that the performance of optimisation techniques can vary depending on the specific characteristics of the dataset and the optimisation problem at hand [1]. Therefore, we should conduct thorough comparative analyses using benchmark datasets and performance metrics relevant to their research objectives to make informed decisions regarding the optimisation technique.

4.3 Previous Applications and Success Stories

The application of ABC algorithms in gene expression studies spans a wide range of research areas, including cancer classification, biomarker discovery, drug response prediction, and pathway analysis [1]. Previous studies have reported notable successes in leveraging ABC algorithms to address various optimisation challenges in these domains. One prominent application of ABC algorithms in gene expression analysis is feature selection, where we aim to identify subsets of genes that are most relevant for discriminating between different disease states or phenotypic conditions. Various findings show that informative gene signatures associated with specific cancer subtypes, drug response phenotypes, and disease prognosis were successfully identified using ABC algorithms for feature selection. Furthermore, ABC algorithms have been utilised to optimise classification models for predicting clinical outcomes or treatment responses based on gene expression profiles [8, 25, 29]. By optimising model parameters and feature subsets using ABC algorithms, studies show improved predictive performance and enhanced interpretability of classification models in diverse biomedical contexts. In addition to feature selection and classification, ABC algorithms have been applied to other optimisation tasks in gene expression analysis, such as clustering analysis and pathway enrichment. Optimising clustering algorithms and identifying robust gene clusters have shown insights into the underlying structure of gene expression data and the functional relationships between co-expressed genes. The application of ABC expression studies has yielded promising results and contributed to advancing our understanding of complex biological processes and disease mechanisms. ABC algorithms drive innovation in biomedical research and personalised medicine by providing efficient and scalable solutions to optimisation challenges [8, 15].

5 Computational Pipeline

5.1 Data Collection

Selection of Datasets: Several criteria must be considered when choosing gene expression research datasets. First, the datasets should be pertinent to the subject or theory under investigation. Studies should consider whether the databases contain samples from the appropriate tissue kinds, environments, and experimental configurations. Furthermore, data quality is critical; datasets with good data quality metrics—like low noise levels and little batch effects—should be given priority. Additionally, it is crucial to ensure that the datasets chosen balance sample size and processing resources while being big enough to offer statistical power for analysis [1].

Data Sources and Integration: Sources for data collection may include in-house experiments, public repositories (such as Gene Expression Omnibus and The Cancer Genome Atlas), partnerships with other research organisations, or other sources, depending on the study's goals. To capture biological variability and increase sample size, it can be required to integrate several datasets from various sources. Data harmonisation techniques, however, need to be carefully considered to handle possible batch effects and guarantee interoperability between datasets [4, 16].

5.2 Preprocessing Steps

Data Cleaning: To eliminate technical artefacts, outliers, and missing values, raw gene expression data frequently has to be preprocessed before analysis can begin. Preprocessing procedures frequently used are background correction, normalisation (robust spline normalisation or quantile normalisation), and batch effect removal methods (surrogate variable analysis and ComBat) [27, 36].

Feature Selection and Dimensionality Reduction: Analysing and interpreting high-dimensional gene expression data can be difficult. To reduce the dimensionality of the data and find pertinent features or patterns, feature selection techniques like differential expression analysis (e.g., limma, DESeq2) or dimensionality reduction techniques like principal component analysis (PCA) or t-distributed stochastic neighbour embedding (t-SNE) can be used [9, 12].

Data Transformation: Techniques like log transformation or variance stabilising transformation may be used to stabilise the variance and guarantee that the data satisfies the presumptions of downstream analysis algorithms [44, 58].

5.3 Implementation Details of ABC Algorithm

The initialisation of the ABC Algorithm: Several scout bees, maximum generations, convergence criteria, and population size are among the initialisation parameters needed for the ABC algorithm. The features of the problem, the available computing power, and the required convergence speed must all be considered while selecting these factors [44].

Objective Function Definition: The fitness landscape of the ABC algorithm seeks to maximise, represented by the objective function in the context of gene expression investigations [59]. Depending on the research issue, the aim function may include finding differentially expressed genes, grouping gene expression profiles, or optimising feature selection criteria.

Parameter Tuning and Optimisation: To achieve the best possible performance of the ABC algorithm on gene expression data, it is essential to fine-tune algorithmic parameters, such as the employed bee types (e.g., employed bees, onlooker bees, scout bees), exploration and exploitation rates, and neighbourhood search strategies [53].

Integration with Biological Knowledge: It is recommended that domain-specific biological knowledge be incorporated into the optimisation process to improve the biological relevance of the outcomes produced by the ABC algorithm [18]. This could include earlier data on pathway enrichment analysis, gene–gene relationships, or functional annotation of gene sets.

Validation and Performance Evaluation: Lastly, using suitable validation metrics, such as clustering validity indices, classification accuracy, or biological relevance measures, the ABC algorithm's performance should be thoroughly evaluated. Permutation testing and cross-validation can be used to evaluate the findings' robustness and generalisability. Furthermore, contrasting the ABC strategy for gene expression analysis with baseline techniques or other optimisation algorithms can shed light on its success [5].

6 Case Studies

We represent the previous work on gene expression profiles using the ABC algorithm, modified for the gene selection procedure for the cancer classification, experimented with by Moosa et al. [34]. This study has a significant contribution to medical diagnosis and further research.

6.1 *Objective and Scope of the Case Study*

This case study aims to explore the effectiveness of ABC algorithms in gene expression dataset analysis to identify patterns and potential biomarkers linked to certain diseases or phenotypes [34]. The scope includes applying ABC algorithms to analyse gene expression data, emphasising how well they perform feature selection, classification, and clustering tasks. The study aims to provide insights into how well ABC algorithms work in gene expression investigations compared to more conventional methods and highlight their advantages and disadvantages.

Microarray data has become a significant subject of extraordinary interest in advancing biologically applicable models from gene expression information in bioinformatics, oncology, and clinical genetics. In this case study, we represent the modified ABC for the gene selection in cancer classification. Only a modest number of gene expression data is compared to the entire number of genes with a critical relationship with specific phenotypes. From the gene selection procedure, we can get significant information from the hereditary ideas of the disease and the process responsible for it. In addition, this procedure improves the performance of cancer classification and can reduce the cost and time of medical diagnosis and prognosis Moosa et al. [34]. This study represents a modified ABC algorithm which selects only a minimum number of genes considered noteworthy for cancer and improves prognosis accuracy. ABC search equation is considered suitable only for the examination but is poorly manipulated. To avoid these problems, this study presents a modified ABC algorithm by associating it with the concept of pheromones, one of the best components of Ant colony optimisation, in which successive bees share their findings through communication.

Table 1 The Characteristics of the dataset used in research validation

Dataset	Sample-size	No. of genes	References
9_Tumors	60	5726	[51]
11_Tumors	174	12,533	[53]
Brain_Tumor1	90	5920	[38]
Brain_Tumor2	50	10,367	[35]
DLBCL	77	5469	[47]
Leukaemia1	72	5327	[20]
Leukaemia2	72	11,225	[7]
Lung Cancer	203	12,600	[10]
Prostate Tumour	102	10,509	[50]
SRBCT	83	2308	[28]

6.2 Selection of Gene Expression Dataset

A gene expression dataset associated with a specific disease or trait is chosen for this case study. To enable comparison analysis, the dataset should include gene expression profiles of samples from both healthy and sick stages. Additionally, the dataset must be sufficiently big and varied to reasonably assess the ABC algorithm's performance. For example, the UCI Machine Learning Repository's Breast Cancer Wisconsin (Diagnostic) Dataset contains gene expression information from breast tissue samples categorised as benign or malignant [34].

This case study is a publicly available gene expression dataset, which accommodates both binary and multiclass high-dimensional data and has been considered for cancer classification to perform the modified ABC method available at http://www.gems-system.org. Datasets are allocated in MATLAB data files. Every file accommodates a matrix, genes, and diagnosis samples. The dataset's attributes utilise all experimental validation, as shown in Table 1.

6.3 Experimental Setup and Parameters

The gene expression dataset is preprocessed in the experimental setup to ensure it is suitable for analysis. This usually involves removing noise that was added during the data-collecting process, normalising expression levels to take sample variances into account, and managing missing values to prevent bias in later studies. The data is clean, standardised, and ready to be analysed through preprocessing. The ABC algorithm is used to carry out tasks like feature selection, classification, or clustering once the dataset has undergone preprocessing. Feature selection aims to identify a group of genes that are most important for distinguishing between various phenotypes or conditions. While clustering tries to organise samples into discrete groups based on similarities in their expression patterns, classification includes assigning samples to

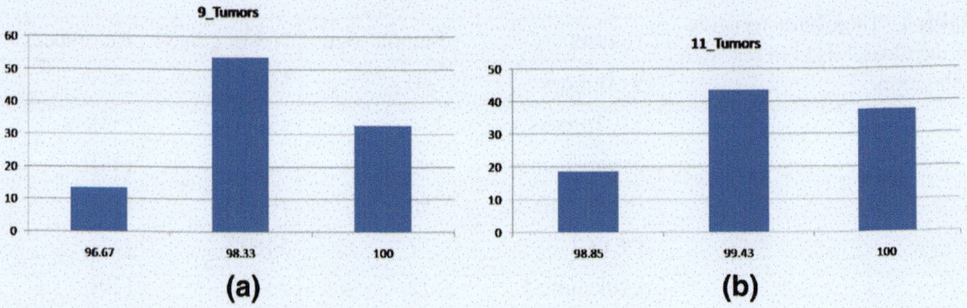

Fig. 4 Distribution of classification accuracy using first (optimised) parameter setting for the dataset (**a**) 9_Tumors; (**b**) 11_Tumors (*Image Credit* [34], doi: https://doi.org/10.1186/s12920-016–0204-7)

predetermined categories (e.g., normal vs. diseased) based on their gene expression profiles [34].

The performance of the ABC algorithm is largely dependent on its parameters. These variables include the colony size of the bees, the maximum number of iterations, the convergence criteria (such as a stopping criterion based on fitness improvement), and the trade-off between exploration and exploitation (e.g., balancing the discovery of novel solutions with the exploitation of promising ones). Certain parameters must be carefully adjusted for the ABC algorithm to operate at its best. The best parameter values that maximise performance on the provided dataset can be found using grid search or cross-validation strategies. Parameter settings were selected based on significant factors rather than the performance obtained. After deliberating the results, the limit is set to 35 for parameter setting.

The algorithm was performed constantly for all datasets based on the number of selected genes and standard deviation for accuracy for optimised setting, and the algorithm has acquired valid accuracy for both default and optimised parameter settings with a high standard deviation in most cases for the number of selected genes in their case study. This method acquired 100% accuracy for all selected datasets except 9_Tumor and 11_Tumor, as shown in Fig. 4 [34].

Achieving the best parameter settings for tuning parameters with complete factorial combination has been done but requires much computational time.

6.4 Evaluation Metrics

Algorithm performance in gene expression research is usually assessed using metrics that measure how well the algorithms extract biological insights from the data. The particular task, such as feature selection, classification, or clustering, determines the evaluation metrics for use.

For Feature Selection: The effectiveness of the chosen features in differentiating between various phenotypes is assessed using metrics like F1-score, area under the receiver operating characteristic curve (AUC-ROC), and classification accuracy [8, 58].

For Classification or Clustering: The ABC algorithm's overall performance in classifying samples or clustering gene expression profiles is evaluated using accuracy, precision, recall, F1-score, and silhouette coefficient metrics [8, 58].

6.5 Comparative Analysis

The performance of the ABC algorithm is compared to baseline techniques frequently employed in gene expression research as part of the comparative study. This comparison aims to see whether the ABC algorithm has any advantages over conventional methods regarding accuracy, efficiency, or resilience. Baseline techniques, like Random Forests, k-means clustering, and Support Vector Machines (SVM), may include those frequently employed in gene expression research. Statistical tests like t-tests and Wilcoxon signed-rank tests may be performed to ascertain the significance of performance differences [50]. Statistical tests can be used to compare the ABC algorithm's performance to baseline techniques and assess whether any reported performance changes are statistically significant. T-tests and Wilcoxon signed-rank tests are popular methods for determining if observed differences are unlikely to result from chance. We can make well-informed decisions regarding the appropriateness of the ABC algorithm for various datasets and applications by utilising comparative analysis, which offers insightful information about the relative advantages and disadvantages of the algorithm in gene expression studies [50].

6.6 Insights Gained from the Case Study

The comparative analysis provides insights into the advantages and disadvantages of the ABC algorithm in gene expression research. These insights may include the ABC algorithm's proficiency in managing high-dimensional gene expression data and locating pertinent features, how parameter settings affect the computational efficiency and algorithm performance, the applicability of results to various diseases and gene expression datasets, and a practical approach to integrating ABC algorithms into gene expression analysis workflows in the real world. By conducting this case study, we can gain valuable insights into computational biology and personalised medicine while gaining important insights regarding the suitability of ABC algorithms in gene expression studies [34, 50].

7 Challenges and Future Research Directions

Despite the promise of ABC algorithms in gene expression studies, several challenges are encountered during their application:

High Dimensionality: Gene expression datasets often contain thousands of genes, leading to high-dimensional optimisation problems. ABC algorithms may struggle to explore such vast solution spaces efficiently, requiring sophisticated strategies to enhance search efficiency [60].

Parameter Tuning: ABC algorithms set parameters such as colony size, limit, and neighbourhood size. Choosing appropriate parameter values can significantly impact algorithm performance but often requires manual tuning, which can be time-consuming and resource-intensive [60].

Handling Noisy Data: Gene expression data is prone to noise from various sources, including experimental variability and technical artefacts. ABC algorithms need to be robust enough to distinguish signal from noise and effectively extract meaningful patterns amidst the noise [22].

Complexity of Biological Systems: Gene regulation networks are highly complex, with various factors influencing interactions between genes and proteins. Capturing the intricacies of biological systems in optimisation models presents a significant challenge for ABC algorithms, requiring advancements in modelling techniques [22].

Interpretability of Results: While ABC algorithms can identify informative gene subsets, interpreting the biological relevance of these results remains challenging. Integrating domain knowledge and biological context into the optimisation process is crucial for ensuring the biological validity of discovered patterns [60].

Scalability: As gene expression studies increasingly utilise large-scale datasets, scalability becomes critical for ABC algorithms. Ensuring scalability to handle massive datasets while maintaining computational efficiency is essential for practical applications.

Addressing the challenges outlined above presents exciting opportunities for future research in applying ABC algorithms to gene expression studies. Here are a few listed below-

Algorithmic Enhancements: Developing advanced ABC variants tailored to the specific characteristics of gene expression data can improve optimisation performance. It includes incorporating adaptive mechanisms, hybridising with other optimisation techniques, and devising strategies for handling high-dimensionality and noisy data [22].

Integration of Prior Knowledge: Integrating prior biological knowledge and domain expertise into ABC algorithms can enhance their interpretability and biological relevance. Incorporating biological constraints and network information can guide the optimisation process towards biologically meaningful solutions [1, 22].

Multi-objective Optimisation: Gene expression analysis often involves multiple conflicting objectives, such as maximising classification accuracy while minimising feature redundancy. Multi-objective optimisation techniques integrated with ABC algorithms can optimise multiple objectives, providing a more comprehensive understanding of gene expression patterns [1, 32].

Scalable Solutions: Developing scalable ABC algorithms that efficiently handle large-scale gene expression datasets is essential for real-world applications. Parallelisation, distributed computing, and optimisation of computational resources can enhance scalability without sacrificing performance [32, 45].

Interdisciplinary Collaboration: Fostering collaboration between computational biologists, bioinformaticians, and optimisation experts is crucial for advancing ABC algorithms' applicability in gene expression studies. Leveraging expertise from diverse fields can lead to innovative solutions and holistic approaches to gene expression analysis [27].

Validation and Benchmarking: Robust validation and benchmarking frameworks are essential for evaluating the performance of ABC algorithms in gene expression studies. Standardised datasets, metrics, and experimental protocols enable fair comparisons between different optimisation techniques and facilitate reproducibility and transparency in research outcomes [32, 45].

8 Conclusion

The application of ABC algorithms in gene expression studies represents a promising avenue for unravelling the complexities of biological systems and elucidating the mechanisms underlying gene regulation and disease processes. Throughout the case study, we have explored the rationale for employing ABC algorithms, examined their application in gene expression analysis, and identified challenges and future research directions in this domain. By leveraging the collective intelligence of artificial bees to navigate high-dimensional and noisy solution spaces, ABC algorithms offer a powerful approach for feature selection, classification, clustering, and pathway analysis tasks in gene expression studies. Their ability to balance exploration and exploitation, robustness to noisy data, and adaptability to high dimensionality make them well-suited for addressing the optimisation challenges inherent in analysing genomic datasets. However, several challenges must be addressed to fully harness the potential of ABC algorithms in gene expression studies. These include handling high dimensionality, parameter tuning, managing noisy data, interpreting results in biological contexts, ensuring scalability, and enhancing algorithmic efficiency. Overcoming these challenges requires interdisciplinary collaboration, algorithmic enhancements, prior biological knowledge integration, and scalable solutions development. Integrating ABC algorithms with gene expression studies significantly advances computational biology and personalised medicine. By providing efficient and scalable solutions to optimisation challenges, ABC algorithms enable extracting meaningful insights from genomic datasets, paving the way for novel discoveries and advancements in biomedical research. As we continue to explore the intricacies of gene expression regulation, ABC algorithms stand poised to play a central role in driving innovation and fostering breakthroughs in our understanding of biological systems.

References

1. Ab Wahab MN, Nefti-Meziani S, Atyabi A (2015) A comprehensive review of swarm optimization algorithms. PLoS One 10(5):e0122827
2. Ahmad S, Chitkara P, Khan FN, Kishan A, Alok V, Ramlal A, Mehta S (2021) Mobile technology solution for COVID-19: surveillance and prevention. In: Computational intelligence methods in COVID-19: surveillance, prevention, prediction and diagnosis, pp 79–108
3. Almugren N, Alshamlan H (2019) A survey on hybrid feature selection methods in microarray gene expression data for cancer classification. IEEE Access 7:78533–78548
4. Alshamlan H, Badr G, Alohali Y (2019) Microarray gene selection and cancer classification method using artificial bee colony and SVM algorithms (ABC-SVM). In: Proceedings of the international conference on data engineering 2015 (DaEng-2015)
5. Alshamlan HM, Badr GH, Alohali YA (2015) Genetic Bee Colony (GBC) algorithm: a new gene selection method for microarray cancer classification. Comput Biol Chem 56:49–60
6. Armañanzas R, Inza I, Santana R, Saeys Y, Flores JL, Lozano JA, Peer YVD, Blanco R, Robles V, Bielza R (2008) A review of estimation of distribution algorithms in bioinformatics. BioData Min 1:1–12
7. Armstrong SA, Staunton JE, Silverman LB, Pieters R, den Boer ML, Minden MD, Sallan SE, Lander ES, Golub TR, Korsmeyer SJ (2002) MLL translocations specify a distinct gene expression profile that distinguishes a unique leukemia. Nat Genet 30(1):41–47
8. Bansal JC, Sharma H, Jadon SS (2013) Artificial bee colony algorithm: a survey. Int J Adv Intell Parad 5(1–2):123–159
9. Ben-Dor A, Bruhn L, Friedman N, Nachman I, Schummer M, Yakhini Z (2000) Tissue classification with gene expression profiles. In: Proceedings of the fourth annual international conference on computational molecular biology
10. Bhattacharjee A, Richards WG, Staunton J, Li C, Monti S, Vasa P, Ladd C, Beheshti J, Bueno R, Gillette M (2001) Classification of human lung carcinomas by mRNA expression profiling reveals distinct adenocarcinoma subclasses. Proc Natl Acad Sci 98(24):13790–13795
11. Booker LB, Goldberg DE, Holland JH (1989) Classifier systems and genetic algorithms. Artif Intell 40(1–3):235–282
12. Brajevic I, Tuba M (2013) An upgraded artificial bee colony (ABC) algorithm for constrained optimization problems. J Intell Manuf 24(4):729–740
13. Calvet L, Benito S, Juan AA, Prados F (2023) On the role of metaheuristic optimization in bioinformatics. Int Trans Oper Res 30(6):2909–2944
14. Črepinšek M, Liu S-H, Mernik L (2012) A note on teaching–learning-based optimization algorithm. Inf Sci 212:79–93
15. Darwish A (2018) Bio-inspired computing: algorithms review, deep analysis, and the scope of applications. Futur Comput Inform J 3(2):231–246
16. Das K, Mishra D, Shaw K (2016) A metaheuristic optimization framework for informative gene selection. Inform Med Unlocked 4:10–20
17. Delgado FM, Gómez-Vela F (2019) Computational methods for gene regulatory networks reconstruction and analysis: a review. Artif Intell Med 95:133–145
18. Fister Jr I, Fong S, Brest J, Fister I (2014) A novel hybrid self-adaptive bat algorithm. Sci World J
19. Flavell SW, Greenberg ME (2008) Signaling mechanisms linking neuronal activity to gene expression and plasticity of the nervous system. Annu Rev Neurosci 31:563–590
20. Golub TR, Slonim DK, Tamayo P, Huard C, Gaasenbeek M, Mesirov JP, Coller H, Loh ML, Downing JR, Caligiuri MA (1999) Molecular classification of cancer: class discovery and class prediction by gene expression monitoring. Sci 286(5439):531–537

21. Gupta S (2022) Enhanced harmony search algorithm with non-linear control parameters for global optimization and engineering design problems. Eng Comput 38(4):3539–3562
22. Hameed SS, Hassan WH, Latiff LA, Muhammadsharif FF (2021) A comparative study of nature-inspired metaheuristic algorithms using a three-phase hybrid approach for gene selection and classification in high-dimensional cancer datasets. Soft Comput 25:8683–8701
23. Houssein EH, Oliva D, Samee NA, Mahmoud NF, Emam MM (2023) Liver cancer algorithm: a novel bio-inspired optimizer. Comput Biol Med 165:107389
24. Huerta EB, Duval B, Hao J-K (2008) Gene selection for microarray data by a LDA-Based genetic algorithm. PRIB 5265:250–261
25. Jabeen A, Ahmad N, Raza K (2022) Global gene expression and docking profiling of COVID-19 infection. Front Genet 13:870836
26. Karaboga D, Basturk B (2007) A powerful and efficient algorithm for numerical function optimization: artificial bee colony (ABC) algorithm. J Global Optim 39:459–471
27. Khan FN, Khanam AA, Ramlal A, Ahmad S (2021) A review on predictive systems and data models for covid-19. Computational intelligence methods in COVID-19: surveillance, prevention, prediction and diagnosis, pp 123–164
28. Khan J, Wei JS, Ringner M, Saal LH, Ladanyi M, Westermann F, Berthold F, Schwab M, Antonescu CR, Peterson C (2001) Classification and diagnostic prediction of cancers using gene expression profiling and artificial neural networks. Nat Med 7(6):673–679
29. Koc E (2010) Bees algorithm: theory, improvements and applications. Cardiff University (United Kingdom)
30. Konar M, BAĞIŞ A (2016) Performance comparison of particle swarm optimization, differential evolution and artificial bee colony algorithms for fuzzy modelling of nonlinear systems. Elektronika ir Elektrotechnika 22(5)
31. Mao L, Van Hemert JL, Dash S, Dickerson JA (2009) Arabidopsis gene co-expression network and its functional modules. BMC Bioinf 10:1–24
32. Marguerat S, Wilhelm BT, Bähler J (2008) Next-generation sequencing: applications beyond genomes. In: Portland Press Ltd
33. Molla M, Waddell M, Page D, Shavlik J (2004) Using machine learning to design and interpret gene-expression microarrays. AI Mag 25(1):23–23
34. Moosa JM, Shakur R, Kaykobad M, Rahman MS (2016) Gene selection for cancer classification with the help of bees. BMC Med Genomics 9:135–165
35. Nutt CL, Mani D, Betensky RA, Tamayo P, Cairncross JG, Ladd C, Pohl U, Hartmann C, McLaughlin ME, Batchelor TT (2003) Gene expression-based classification of malignant gliomas correlates better with survival than histological classification. Can Res 63(7):1602–1607
36. Passino KM, Seeley TD (2006) Modeling and analysis of nest-site selection by honeybee swarms: the speed and accuracy trade-off. Behav Ecol Sociobiol 59:427–442
37. Paszkowicz W (2009) Genetic algorithms, a nature-inspired tool: survey of applications in materials science and related fields. Mater Manuf Process 24(2):174–197
38. Pomeroy SL, Tamayo P, Gaasenbeek M, Sturla LM, Angelo M, McLaughlin ME, Kim JY, Goumnerova LC, Black PM, Lau C (2002) Prediction of central nervous system embryonal tumour outcome based on gene expression. Nature 415(6870):436–442
39. Qazi S, Khanam A, Raza K (2022) Potential role of the nature-inspired algorithms for classification of high-dimensional and complex gene expression data. Nature-Inspired intelligent computing techniques in bioinformatics, studies in computational intelligence, vol 1066. Springer, pp 89–102
40. Rana N, Latiff MSA, Abdulhamid SIM, Chiroma H (2020) Whale optimization algorithm: a systematic review of contemporary applications, modifications and developments. Neural Comput Appl 32:16245–16277
41. Raza K (2016) Analysis of microarray data using artificial intelligence based techniques. Handbook of research on computational intelligence applications in bioinformatics. IGI Global, USA, pp 216–239

42. Raza K, Ahmad S (2019) Recent advancement in next-generation sequencing techniques and its computational analysis. Int J Bioinform Res Appl 15(3):191–220
43. Raza K, Hasan AN (2015) A Comprehensive evaluation of machine learning techniques for cancer class prediction based on microarray data. Int J Bioinform Res Appl 11(5):397–416
44. Rice JR (1976) The algorithm selection problem. In: Advances in computers, vol 5. Elsevier, pp 65–118
45. Rihoux B, Ragin CC (2008) Configurational comparative methods: qualitative comparative analysis (QCA) and related techniques. Sage Publications
46. Sahu A, Qazi S, Raza K, Singh A, Verma S (2022) Machine learning-based approach for early diagnosis of breast cancer using biomarkers and gene expression profiles. In: Computational intelligence in oncology, studies in computational intelligence (SCI), vol 1016. Springer, pp 285–306
47. Shipp MA, Ross KN, Tamayo P, Weng AP, Kutok JL, Aguiar RC, Gaasenbeek M, Angelo M, Reich M, Pinkus GS (2002) Diffuse large B-cell lymphoma outcome prediction by gene-expression profiling and supervised machine learning. Nat Med 8(1):68–74
48. Shukla AK, Tripathi D, Reddy BR, Chandramohan D (2020) A study on metaheuristics approaches for gene selection in microarray data: algorithms, applications and open challenges. Evol Intel 13(3):309–329
49. Singh A, Deep K (2019) Exploration–exploitation balance in artificial bee colony algorithm: a critical analysis. Soft Comput 23:9525–9536
50. Singh D, Febbo PG, Ross K, Jackson DG, Manola J, Ladd C, Tamayo P, Renshaw AA, D'Amico AV, Richie JP (2002) Gene expression correlates of clinical prostate cancer behavior. Cancer Cell 1(2):203–209
51. Staunton JE, Slonim DK, Coller HA, Tamayo P, Angelo MJ, Park J, Scherf U, Lee JK, Reinhold WO, Weinstein JN (2001) Chemosensitivity prediction by transcriptional profiling. Proc Natl Acad Sci 98(19):10787–10792
52. Stephanopoulos G, Alper H, Moxley J (2004) Exploiting biological complexity for strain improvement through systems biology. Nat Biotechnol 22(10):1261–1267
53. Su AI, Welsh JB, Sapinoso LM, Kern SG, Dimitrov P, Lapp H, Schultz PG, Powell SM, Moskaluk CA, Frierson HF Jr (2001) Molecular classification of human carcinomas by use of gene expression signatures. Cancer Res 61(20):7388–7393
54. Tang J, Duan H, Lao S (2023) Swarm intelligence algorithms for multiple unmanned aerial vehicles collaboration: a comprehensive review. Artif Intell Rev 56(5):4295–4327
55. Tran B, Xue B, Zhang M (2014) Overview of particle swarm optimisation for feature selection in classification. Simulated evolution and learning: In: 10th international conference, SEAL 2014, Dunedin, New Zealand, 15–18 Dec 2014. Proceedings 10
56. Tripathi MK, Ahmad S, Tyagi R, Dahiya V, Yadav MK (2022) Fundamentals of molecular modeling in drug design. In: Computer Aided Drug Design (CADD): From ligand-based methods to structure-based approaches. Elsevier, 125–155
57. Voll P, Jennings M, Hennen M, Shah N, Bardow A (2015) The optimum is not enough: A near-optimal solution paradigm for energy systems synthesis. Energy 82:446–456
58. Yadav MK, Bhutani K, Ahmad S, Raza K, Singh A, Kumar S (2024) Application of machine learning–based approaches in stem cell research. In: Computational biology for stem cell research. Elsevier, pp 65–76
59. Zhao F, Wang Z, Wang L, Xu T, Zhu N (2023) An exploratory landscape analysis driven artificial bee colony algorithm with maximum entropic epistasis. Appl Soft Comput 137:110139
60. Zhou J, Lilhore UK, Hai T, Simaiya S, Jawawi DNA, Alsekait D, Ahuja S, Biamba C, Hamdi M (2023) Comparative analysis of metaheuristic load balancing algorithms for efficient load balancing in cloud computing. J Cloud Comput 12(1):85

Artificial Bee Colony Algorithm in Multi-omics Analysis: A Case Study

Saira Hamid, Alisha Ansari, and Raiyan Ali

1 Introduction

Multi-omics is a combination and interaction of diverse molecular levels of biological information such as genomics, transcriptomics, epigenomics, proteomics, metabolomics, microbiomes, and others [1]. The integration of these datasets provides an understanding of the relationship between driver genomic alterations, genomic mutations, and protein expression that explores cellular responses to therapy, pathogenic changes, and novel associations between various biomolecules [2]. It plays an important role in understanding complex biological diseases like cancer [3] which creates a need for new therapeutic opportunities. The Single-omics technologies provide valuable data but they offer a limited view of cellular functions that are unable to capture the complete complexity of diseases [4]. In contrast, integrated multi-omics data allows an understanding of causal relationships across various molecular levels. This strategy accounts for high dimensionalities and complexities in data interpretation enabling robust analytical tools to be developed that enhance our understanding of diseases and human physiology [5]. High-throughput sequencing technologies are employed to generate multi-omics data [6] requiring a profound understanding of disease pathogenesis and tumor malignancy mechanisms. Applications like network biology, dimensionality reduction, biomarker identification, and enrichment analysis facilitate valuable insights from multi-omics analysis. Despite its immense potential, adopting multi-omics in clinical settings faces challenges, including the need for standardized analytical pipelines,

S. Hamid
Department of Computer Science, Jamia Millia Islamia, New Delhi, India

A. Ansari
School of Computational and Integrative Sciences, Jawaharlal Nehru University, New Delhi, India

R. Ali (✉)
Department of Biochemistry, All India Institute of Medical Sciences (AIIMS), New Delhi, India
e-mail: raiyanali92@gmail.com

K. Raza (ed.), *Solving with Bees*, Springer Tracts in Nature-Inspired Computing,
https://doi.org/10.1007/978-981-97-7344-2_11

multidisciplinary training, and contributing feature selection. Overcoming these challenges is pivotal for translating practical clinical applications, and contributing to advancements in personalized medicine, diagnostics, and prognosis. The large and complex nature of integrated multi-omics datasets necessitates feature optimization to identify molecular features contributing to the development of accurate diagnostic and prognostic modeling.

1.1 Necessity and Advancement of Feature Optimization

The diverse availability of multi-omics data types poses a challenge in analysis, as a significant portion is non-informative due to redundancy or irrelevance [7]. Optimization addresses the challenge of multitudinous input features and enhances the inductive learner's capabilities and model simplicity. Predictive information often overlaps, varies, and exhibits interactions among features from distinct data types, adding complexity to the analysis [8]. Moreover, the inclusion of clinical covariates in phenotypic datasets emphasizes the need for feature optimization [9]. Several feature selection methods exist, and their application to multi-omics data depends on the complexity of the data and the research question of interest. One benchmarking study compares filter methods, embedded methods, and wrapper methods such as the ant colony algorithm which is inspired by Swarm intelligence, evaluating their performance in selecting feature combinations suitable for effective classification [10]. Additionally, the study assesses the influence of combining multi-omics data with clinical information on classification performance. Swarm Intelligence optimization algorithms are a kind of Nature-inspired algorithms based on nature's processes such as evolution and natural selection that mimic these algorithms on gene expression data with large dimensions giving a better disease classification as well as biomarker determination [11]. The use of these algorithms in bioinformatics has shown great potential for tackling extremely difficult optimization problems thus proving their worth in handling large and complex multi-omics dataset studies [12].

In the era of Advancements in feature optimization methods, hybrid methods are coming into the field due to the complexity of real-time data as seen in biomedical research [10], particularly in navigating the convolutions of genomic data. One remarkable evolution in this domain involves the strategic application of clustering techniques, such as CLARANS and FCLARANS, for addressing feature (gene) selection challenges. CLARINS, a randomized search-based method, extracts feature sets into biologically enriched gene clusters [13], pioneering the integration of biological knowledge into feature extraction for optimized outcomes. The subsequent introduction of FCLARANS [14], featuring fuzzy clustering, signifies a refinement in feature selection techniques, offering detailed novel optimization strategies. One study showcases the ABCD (Artificial Bee Colony based on Dominance) algorithm, a multi-objective approach for selecting optimal gene subsets in tissue sample classification. Through artificial bee colony (ABC) optimization principles, ABCD employs

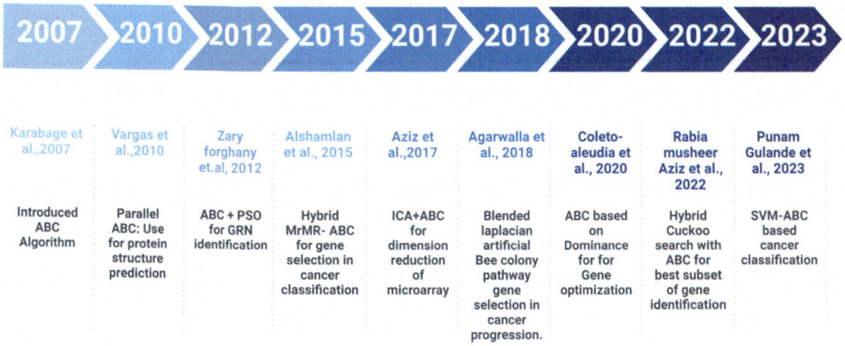

Fig. 1 The evolution of the Artificial Bee Colony (ABC) algorithm, tracing its development from its initial introduction to its application in SVM-ABC-based classification. It illustrates the progressive enhancements and adaptations of the ABC algorithm including its refinement and integration with Support Vector Machines (SVM) to improve classification performance in various applications

a population of artificial bees representing potential gene subsets. These bees iteratively update their positions based on the concept of dominance, balancing the minimization of selected genes and maximization of classification accuracy. Inspired by natural foraging behavior, the algorithm converges toward Pareto front-optimal solutions, effectively addressing the multi-objective optimization problem in gene selection for cancer tissue classification [15]. The ABCD algorithm not only optimizes features but also highlights the significance of balancing conflicting objectives in multi-objective optimization problems. Similarly, enhanced research and the recent development of hybrid methods can be seen in Fig. 1.

This departure from the traditional approach of feature optimization methods emphasizes a researcher-centric approach, allowing for more targeted feature optimization methodologies. This chapter underlined the significance and challenges posed by multi-omics while elucidating why feature optimization is important in multi-omics. Further, it shows the effectiveness and significance of the ABC algorithm, one of the Swarm intelligence-based methods, in optimizing the complex multi-omics datasets through a case study of implementation on real-time data for the selection of informative features. These features can be valuable in the advancement of diagnostic and prognostic modeling.

2 Multi-omics Analysis

2.1 Significance and Challenges

Multi-omics analysis enables the identification of molecular markers linked to biological processes, the discovery of regulatory units, and an understanding of the mechanisms that control cellular fate and molecular phenotypes. This could discover predictive or prognostic biomarkers, facilitate customized medicine development, and identify therapeutic targets, bridging the genotype–phenotype gap. The investigation of information flow across several molecular levels improves the accuracy of prognostics and forecasts in disease phenotypes, improving illness subtyping and biomarker prediction. The integration of various omics datasets, including ChIP-Seq, RNA-Seq, and proteomics data, provides valuable insights into biological systems and prioritizes driver genes, as demonstrated in diseases such as head and neck squamous cell carcinoma and colorectal cancers [16, 17]. The introduction of high-throughput technology has made it possible to collect multi-omics patient samples and perform integrated analysis, which is critical for translational medicine projects. The choice of omics types and integration methodologies is crucial in the design of multi-omics research in translational medicine [18]. However, there is still no agreement on the best combination of omics to include as well as the data integration methodologies needed for their analysis, highlighting the need for additional research in this area.

Complex nonlinearities make it difficult to analyze detailed correlations in multi-omics data, thus complicating the interpretation of underlying patterns. The processing of datasets of varying sizes and missing information presents obstacles to achieving thorough knowledge, affecting the dependability of the analytical results. The management of differences in statistical distributions and data formats adds complexity and requires careful attention to ensure accurate integration and interpretation. The large number of variables in multi-omics datasets relative to sample sizes adds complexity, necessitating the reduction of variables to core features for a meaningful understanding of illness causation. The integration of omics measurements lacks consensus on ideal approaches, posing the challenge of efficiently combining disparate data types from patient samples. Combining patterns with multi-omics analysis in downstream processes improves interpretation but complicates the formulation of specific research hypotheses. Combining trials from various facilities presents challenges that limit the meaningful integration of single-omic datasets. Technical and biological variables lead to heterogeneity in multi-omic datasets, complicating interpretation and failing to meet expectations for uncovering novel biological information. For correct results, machine learning in multi-omics analysis must be applied to a more systematic, precise, and large biological dataset. Omics analyzes face problems in data integration, necessitating careful consideration of factors, such as grouping, visualization, and functional characterization. The growing complexity of the software used in multi-omics analysis presents issues in ensuring repeatability and benchmarking computational tools.

2.2 *Methods and Techniques*

Multi-omics analysis involves the integration of diverse biological datasets to gain an integrated understanding of complex biological systems [19]. Various methodologies and techniques have been developed to extract meaningful insights from multi-omics data, allowing it to uncover hidden patterns, identify biomarkers, and reveal underlying mechanisms in diseases. However, multi-faceted data from different genetic spaces in a cellular organization poses challenges due to differences in dynamic ranges and data distributions. The integration of machine learning (ML) and deep learning (DL) approaches is a crucial aspect of multi-omics analysis. Integration algorithms methods such as similarity-based integration, fusion-based integration, network-based integration, Bayesian-based stochastic integration, correlation-based integration, multivariate methods, and other multimodal concatenation-based techniques are used, and many tools were created based on their question of interest [17].

Similarity-based methods, like Similarity Network Fusion (SNF), construct individual networks per omics and iteratively update them to converge into a single network [20]. SNF helps in identifying cancer subtypes based on inter-patient similarities but does not contribute to biomarker discovery. One prominent category of methods is unsupervised multivariate approaches, such as Joint non-negative matrix factorization (NMF) and Joint and Individual Variation Explained (JIVE) [21]. Joint NMF decomposes non-negative matrices, projecting diverse data types onto a common coordinate system.

Bayesian frameworks, like iCluster, offer a flexible approach for integrating multiple genomic characteristics without non-negative constraints. iCluster employs a Gaussian latent variable model to capture correlative structures in multi-omics data, leading to joint clustering of samples and identification of cluster-relevant features [22]. Another upgraded version, iCluster+, accommodates various data types and has demonstrated accurate classification for simulated datasets. One of the Network-based approaches, exemplified by PAthway Representation and Analysis by Direct Reference to Graphical Models (PARADIGM), utilizes probabilistic graphical models to incorporate curated pathway interactions among genes [23]. PARADIGM successfully identifies altered activities in cancer-related pathways, offering insights into pathway perturbations and clinically relevant subgroups.

Canonical correlation analysis (CCA) based on correlation-based integration assesses correlation across copy number variation, methylation, and gene expression data, providing insights into the mechanisms of carcinogenesis [24]. CCA selects discriminative features from multi-omics data sources, enhancing predictive power in predicting survival outcomes. A comprehensive view of integration methods is shown in Table 1.

The future of multi-omics data integration is promising, with the development of various computing frameworks and tools. As technology advances, more promising insights into complex biological phenomena are expected to gain easier access to multi-omics analyses, unlocking valuable biological insights from integrated datasets.

Table 1 Represents the multi-omics data integration algorithms consisting of distinct methodologies as network approaches, fusion approaches, Bayesian approaches, correlation-based canonical approaches, and other multivariate approaches

Name	Method	Computational platform	References
Joint NMF	Matrix factorization	Python	[24]
SMF	Matrix factorization	Python	[25]
iCluster+	Matrix factorization	R	[26]
iClusterBayes	Matrix factorization	R	[27]
moCluster	Matrix factorization	R	[28]
JIVE	Matrix factorization	MATLAB	[21]
MOFA	PCA	R/Python	[29]
NetICS	Network-based	MATLAB	[13]
BCC	Bayesian	R	[30]
MDI	Bayesian	MATLAB	[31]
PARADIGM	Pathway networks, Bayesian	Python	[23]
iBAG	Multi-step analysis	R	[32]
SNF	Network-based, similarity-based	R/MATLAB	[20]
iOmicsPASS	Network-based	R	[33]
NEMO	Similarity-based clustering	R	[34]
PFA	Fusion-based integration	MATLAB	[35]
CCA	Correlation-based	R	[36]
LRAcluster	Low-rank approximation	R	[37]
COCA	Multivariate clustering	R/Python	[38]
WGCNA	Correlation-based network	R	[39]
DSSF (Deep Subspace Similarity Fusion)	Kernel-based	R/MATLAB	[40]
AFN (Affinity Network Fusion)	Kernel-based	R	[41]

3 Application of Multi-omics in Healthcare

Advances in technology have allowed us to combine knowledge from various fields, such as genomics, transcriptomics, proteomics, and metabolomics, which has helped us bridge the gap between health and disease. For example, in longitudinal sampling using multi-omics, the progression of a disease can be studied over some time. This also helps us study heterogeneity among individuals. This approach enables deep phenotyping, which involves capturing detailed information regarding an individual's characteristics and molecular profiles [6].

Precision health can help predict, prevent, and even cure diseases more precisely by taking into account an individual's genetics and environment, including lifestyle factors. Traditional medicine usually follows the principle of 'one-size-fits-all' and the symptomatic approach, whereas precision health considers the unique characteristics of every individual. Precision health can lead to more effective treatment strategies by incorporating personalized subtyping using precise molecular profiles. Thus, this approach has the potential to improve healthcare efficiency and quality by tailoring interventions according to individual needs. Multi-omics has various benefits for precision health. Different omics technologies provide complementary information, enabling a more comprehensive understanding of molecular complexity. For example, proteomics can provide information about protein expression levels, whereas metabolomics can reveal changes in small-molecule quantities. By merging these forms of data, researchers can acquire a more complete picture of the biological pathways underlying diseases. Applications of multi-omics in healthcare have been shown in Fig. 2.

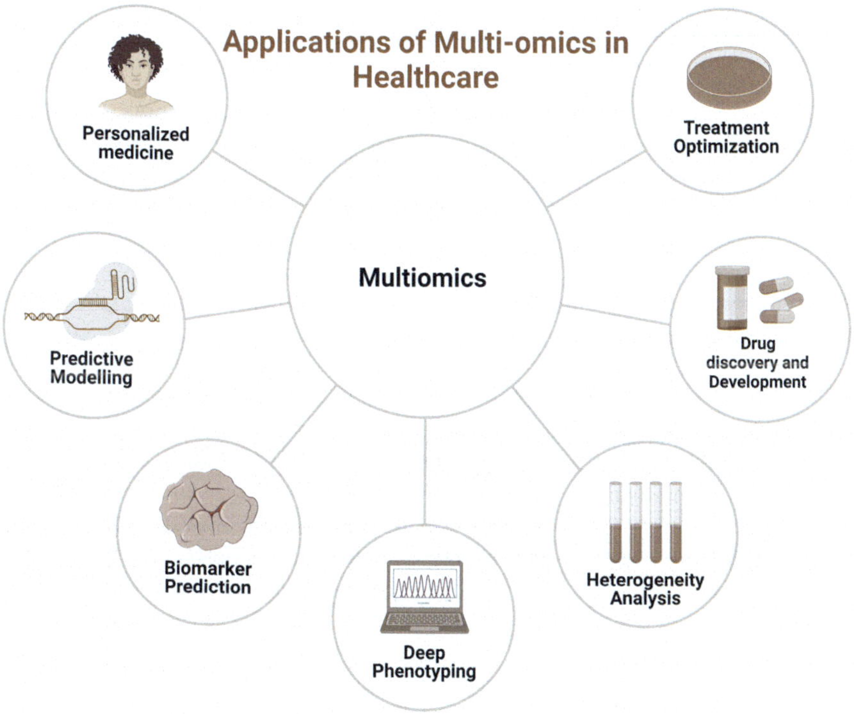

Fig. 2 Applications of multi-omics in healthcare and personalized medicine

Multi-omics data integration analysis has enabled the creation of biomarkers for early detection and tailored therapy. A comprehensive study was done using metabolomics, lipidomics, glycomics, and metallomics methods to create intricate multi-omics maps of interrelated biomolecules, considerably improving our understanding of myocardial infarction [42]. Their findings not only revealed the possible function of specific biomolecules in immune control but also highlighted the importance of certain molecules in cardiovascular disease. They also discovered how certain biomolecules affect the function of essential molecules in the context of illness. Furthermore, multi-omics integration not only aids in the discovery of complicated molecular mechanisms but also greatly promotes the advancement of precision medicine and personalized treatment approaches [43]. Deeper insights into how medications work in treating myocardial infarction can be gained by evaluating data across many omics layers, allowing the development of more tailored and precise treatment methods, optimizing treatment effectiveness, and reducing the likelihood of adverse responses.

4 ABC Algorithm in Multi-omics: Case Study

4.1 Objective and Scope of the Case Study

The ABC algorithm is a nature-inspired optimization technique that operates iteratively, balancing exploration and exploitation to converge toward optimal solutions of complex search spaces. Its core principles, derived from the foraging behavior of honeybee swarms [44], include self-organization, where driving mechanisms lead to global structures through local interactions, incorporating positive and negative feedback, fluctuations, and multiple interactions. Swarm intelligence is such designed algorithms inspired by collaborative behaviors observed in various social entities, such as social insects like bees [45].

Practically, the ABC algorithm structures its bee colony into employed bees, onlookers, and scouts. Employed bees explore and evaluate solutions, onlookers probabilistically select solutions based on fitness values, and scouts introduce diversity when improvements stagnate; implementation steps can be seen in Fig. 3. This structured division ensures an iterative process that efficiently navigates multidimensional and multimodal optimization challenges, maintaining a delicate balance between exploration and exploitation. The objective is to underline the importance of ABC in the feature optimization process of large and complex multi-omics dataset studies.

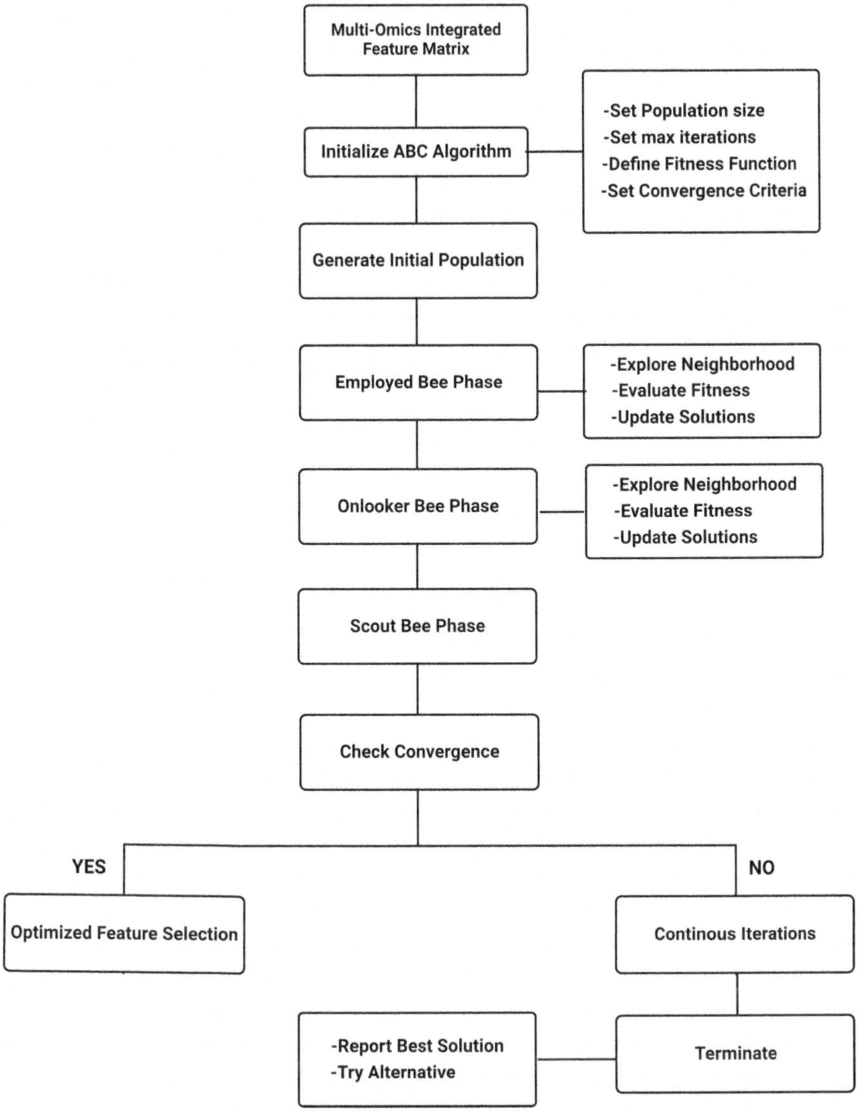

Fig. 3 Workflow illustrating practical implementation of ABC algorithm for feature optimization

4.2 ABC's Role in Biomarker Selection

The ABC algorithm plays a strategic role in optimizing the selection process, standing out for its inherent advantages, such as simplified parameter configuration and faster convergence compared to other evolutionary algorithms [46]. It offers results comparable to or better than alternative evolutionary algorithms, maintaining higher population diversity to prevent premature convergence. The exploration of DNA microarrays using the ABC algorithm with artificial neural networks, as the proposed methodology, provides a broad understanding of ABC's role in optimizing features for enhanced biomarker selection [47]. ABC functions as a metaheuristic algorithm within the Co-ABC framework, refining the gene subset post the Correlation-based Feature Selection (CFS) phase. CFS identifies genes with robust correlations to specific cancer classes and minimal inter-gene correlations [48]. Integrated into the Co-ABC algorithm, ABC ensures a streamlined and effective biomarker selection process, balancing computational efficiency with the selection of relevant and informative genes, marking ABC's significance in medical research and diagnostics. The methodology for multi-omics analysis with ABC can be seen in Fig. 4.

ABC's primary goal is to pinpoint the optimal combination of genes, significantly boosting biomarker performance. Its simplicity and ease of implementation contribute to the selection of a concise, highly predictive gene set. In essence, ABC's decisive role in biomarker selection harmonizes computational efficiency with the identification of informative genes for precise disease categorization.

4.3 *Insights Gained from Case Study*

Breast cancer poses a serious healthcare challenge, with high mortality rates making the development of accurate survival prediction models critical for improving patient outcomes. This case study investigates the proposed parallel Bayesian hyperparameter-optimized Stacked ensemble model, termed BSense, which aims to develop a precise prognostic model for breast cancer survival prediction by integrating multi-omics data. The BSense model leverages various data dimensions, including genomic, proteomic, transcriptomic, metabolomic, and pathological image data, to capture the complex biological processes involved in breast cancer. The ABC algorithm plays a decisive role in feature selection and model optimization for the BSense model. The ABC algorithm is employed to optimize the feature space by selecting the most relevant multi-omics features that are highly correlated with the class label (breast cancer survival). Therefore, it addresses the computational challenges associated with analyzing high-dimensional multi-omics data by efficiently exploring the search space and identifying optimal feature subsets. Additionally, the ABC algorithm refines the model's performance by tuning hyperparameters and adjusting parameters and, as a result, enhances the computational efficiency of the BSense model. The BSense model combines Deep Neural Networks (DNNs), Gradient Boosting Machines (GBMs), and Distributed Random Forests (DRFs) through parallel Bayesian optimization, exploiting the ABC algorithm's efficiency in feature selection. The individual model hyperparameters are tuned using this approach, resulting in an ensemble model that leverages the strengths of each component.

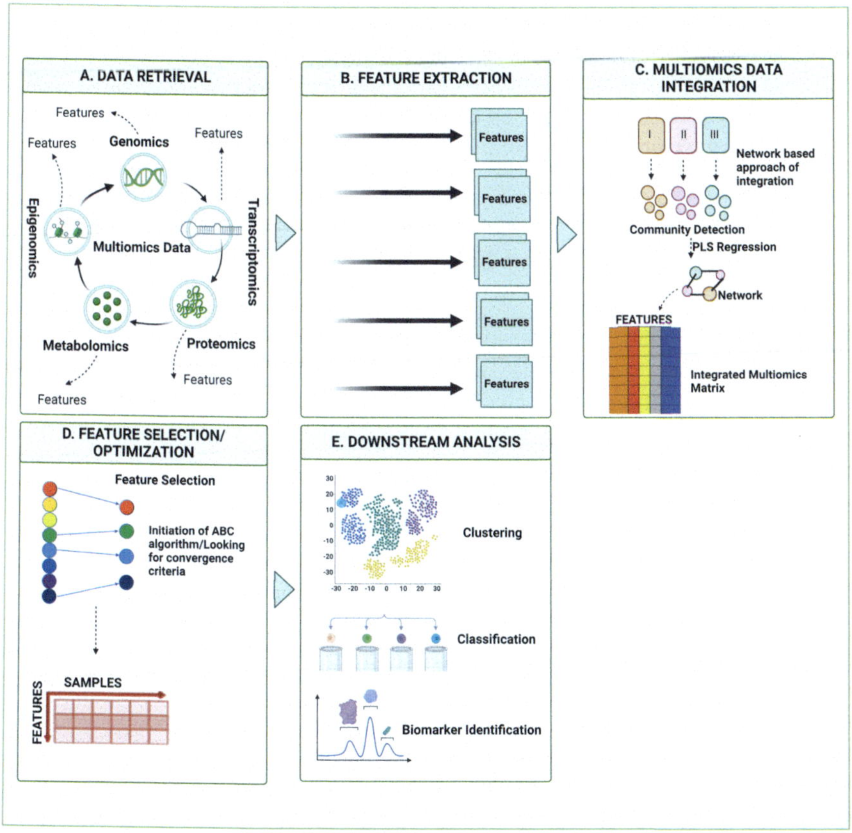

Fig. 4 The implementation of the ABC algorithm in multi-omics analysis. The workflow starts with data extraction (TCGA, cBioPortal, and GTEx), feature extraction using techniques like (PCA, ICA, Sparse methods, and AutoEncoders), multi-omics data integration (horizontal, canonical, and vertical), feature selection and optimization using the ABC algorithm to enhance data quality and downstream analysis which includes classification of samples, clustering to identify new subtypes, and biomarker identification to find specific indicators of biological conditions or diseases

The study evaluates the BSense model's performance on several datasets, including TCGA, METABRIC, Metabolomics, and RNA-seq data. The model achieves impressive AUC estimates of 83.9%, 87.3%, 91.1%, and 80.1% respectively, demonstrating superior prediction rates compared to traditional machine learning algorithms [49]. These results highlight the potential of the ABC algorithm in optimizing feature selection and improving the accuracy of predictive models for breast cancer survival prediction using multi-omics data. It demonstrates the ABC algorithm's effectiveness in selecting the most relevant multi-omics features, improving the accuracy of the proposed BSense model for breast cancer survival prediction. The results show that the BSense model outperforms individual models, providing a comprehensive understanding of the complex biological processes involved in breast cancer. The findings highlight the potential of the ABC algorithm in optimizing feature selection and improving the accuracy of predictive models for breast cancer survival. This achievement not only reaffirms the applicability of the ABC algorithm for optimizing feature selection but also solidifies the BSense model as a promising advancement in the field of cancer prediction models.

5 Current Challenges and Future Prospects

In the dynamically changing landscape of multi-omics analysis, the utilization of optimization algorithms, such as the ABC algorithm, confronts a series of formidable challenges. One of the primary challenges facing the application ABC algorithm in multi-omics analysis is the inherent high dimensionality of the datasets. This high dimensionality is characterized by a multitude of features, and it stands as a formidable obstacle for the ABC algorithm. The complexity arises from the vast number of variables or characteristics present in the data, making it intricate and challenging for the algorithm to navigate and effectively optimize solutions within the multi-faceted space [50]. Moreover, the integration of diverse data types—encompassing genomics, transcriptomics, proteomics, and metabolomics—demands a tailored adaptation of the ABC algorithm to adeptly handle the heterogeneity of these sources, facilitating the merging of information from varied origins [51]. The optimization process faces a critical issue in the form of optimal parameterization, where the sensitivity of the ABC algorithm to parameter settings necessitates meticulous tuning, a task that can be computationally expensive, particularly in the context of multi-omics data [52]. The intricate and nonlinear relationships characteristic of biological systems further challenge the ABC algorithm's ability to converge to optimal solutions within the complex landscapes shaped by these dynamics. The algorithm's scalability emerges as a pivotal concern, requiring careful consideration to ensure its adept scaling with the increasing size of multi-omics datasets, particularly when dealing with real-world complexities [53]. Additionally, addressing the robustness of the ABC algorithm to noise and its capacity to handle missing data becomes imperative in the context of biological data, which often exhibits noise and missing values [54].

As multi-omics analyses aim not only for accuracy but also interpretability, ongoing efforts are essential to enhance the ABC algorithm's capability to deliver results that are not only precise but also comprehensible within the intricate biological context [15]. Finally, establishing robust benchmark datasets and tailored validation metrics specific to multi-omics analysis becomes crucial, providing a foundation for the comprehensive evaluation of the ABC algorithm's performance compared to other optimization techniques [54]. This ongoing refinement contributes significantly to the algorithm's adaptability in the ever-evolving field of multi-omics research.

The prospects of the ABC algorithm in multi-omics analysis are promising, driven by ongoing advancements in optimization techniques and the evolving landscape of biological research [55]. Efforts may focus on refining the algorithm's adaptability to high-dimensional data through improved exploration–exploitation balance and feature selection mechanisms [56]. Integration with advanced machine learning techniques and hybridization approaches could enhance its robustness and efficiency. Dynamic parameter adaptation strategies may be developed to adjust to the unique characteristics of diverse multi-omics datasets [57]. Addressing the challenge of integrating heterogeneous data types effectively and improving scalability for large-scale datasets are crucial directions for future research. The ABC algorithm may find applications in personalized medicine, tailoring treatment strategies based on individual multi-omics profiles [15]. Additionally, efforts to enhance interpretability and explainability, integration with domain-specific knowledge, and the establishment of standardized benchmarking frameworks will contribute to its continued success in advancing our understanding of complex biological systems. Collaboration between computational scientists and biologists is essential for developing more effective tools for multi-omics research.

6 Discussion

With the advancements in high-throughput technologies, it becomes easier to sequence and generate a large quantity of multi-omics data which results in the development of databases day by day consisting of Big data. Enormous information available within this large crunch of biological data garnered the complexity in analysis, integration, and optimization level tasks. This gave rise to several opportunities and challenges in understanding the inner behavior of complex human diseases ranging from cancers to neurological disorders and the ever-evolving viral anomalies. Multi-omics Analysis holds relatively high importance in discerning the disease occurring or genetically present molecular level interactions than through the single-level omics analysis. While being challenged with many steps such as feature extraction, data integration, feature optimization, and identification of biomarkers or their interactive partners, the integrative information of multi-omics datasets provides paralleled insights into disease pathogenesis, tumor classification, prognostic predictions, and identification of therapeutic possibilities.

Several methodologies were developed and are still ongoing in tackling these challenges for integration purposes; network approaches, Bayesian approaches, and many multivariate approaches are still developing with the latest development technologies. While feature optimization is a necessity for diagnostic and prognostic modeling purposes, so many contributions were made from filters to wrapper methods, and also the ever-evolving hybrid methods. Many hybrid algorithms were developed for multi-omics feature optimization purposes as ABC combined with Particle swarm optimization (PSO) [59], Hybrid cuckoo search with ABC [60], correlation-based ABC, and a combination of marine algorithms with Bayesian inferences. Nature-inspired algorithms show many beneficial applications in feature optimization due to their nature of searching and identifying complex patterns. Our chapter focuses on the ABC algorithm which is a swarm intelligence algorithm, a kind of nature-inspired method. It is used to handle high dimensionalities found in multi-omics datasets. In order to tackle the issue of redundancy, and heterogeneous data, complex biological patterns are examined through its feature optimization. The algorithm offers efficient exploration and exploitation, global search capabilities, adaptability to different problem domains, efficient convergence, scalability, robustness, and enhanced classification performance [58]. The aim is to increase the performance of machine learning models by lowering their dimension, increasing their accuracy, and making them learn quickly from the available data by identifying the most relevant and discriminatory features for training purposes. The ABC algorithm in the context of a multi-omics framework has been demonstrated in the above case study, showing its ability to refine feature spaces and optimize the prediction model BSense. The findings show the use of ABC algorithms could redefine cancer predictive modeling and give promising results in the future of personalized medicine, diagnostics, and prognostic modeling.

Regardless of the advancements made in multi-omics analysis, there are still some challenges across it such as standardized analytical pipelines, multidisciplinary training, and dealing with high-dimensionality datasets. These challenges can be overcome through better exploration–exploitation balance, feature selection mechanisms, and scalability for large-scale datasets. Further exploration with better algorithms and technologies would lay the foundation for the efficient working of healthcare.

7 Conclusion

The evolving landscape of multi-omics analysis highlights its significance in advancing the understanding of diseases. The integration of distinct omics datasets provides a comprehensive view of molecular-level interactions and presents novel insights. The importance of feature optimization is addressed through various feature selection methods. It reduces the complexities of high-dimensional datasets of multi-omics, where the ABC algorithm found potential algorithm in optimizing features, to confer biomarker selection. The case study on breast cancer survival prediction demonstrates the practical application of the ABC algorithm, highlights its efficacy in refining feature spaces, and contributes to the more accurate development of predictive models. While being promising in the optimization of complex feature spaces, ABC has also been coupled with many evolutionary algorithms. It can be considered that the hybrid version of evolving ABC can address the challenges in feature optimization of biological data more accurately. Also, the standardization of multi-omics workflow will be developed as a breakthrough through the continuous collaborative efforts of experts. With the advancements of technology, our understanding of complex biological systems will deepen and, coupled with advanced feature optimization methods, will play a promising role in reshaping medical research and development of personalized healthcare.

References

1. Kreitmaier P, Katsoula G, Zeggini E (2022) Insights from multi-omics integration in complex disease primary tissues. Trends Genet
2. Gross AL et al (2020) Derivation of a measure of physiological multisystem dysregulation: results from WHAS and health ABC. Mech Ageing Dev 188:111258–111258
3. Alyass A, Turcotte M, Meyer D (2015) From big data analysis to personalized medicine for all: challenges and opportunities. BMC Med Genomics 8(1)
4. Menyhárt O, Győrffy B (2021) Multi-omics approaches in cancer research with applications in tumor subtyping, prognosis, and diagnosis. Comput Struct Biotechnol J 19:949–960
5. Wani N, Raza K (2019) Integrative approaches to reconstruct regulatory networks from multi-omics data: a review of state-of-the-art methods. Comput Biol Chem 83:107120
6. Chen C et al (2023) Applications of multi-omics analysis in human diseases. MedComm 4(4):e315–e315
7. Gidskehaug L et al (2007) A framework for significance analysis of gene expression data using dimension reduction methods. BMC Bioinform 8(1)
8. Zhao Q et al (2014) Combining multidimensional genomic measurements for predicting cancer prognosis: observations from TCGA. Brief Bioinform 16(2):291–303
9. Herrmann M et al (2020) Large-scale benchmark study of survival prediction methods using multi-omics data. Brief Bioinform 22(3)
10. Li Y et al (2022) Benchmark study of feature selection strategies for multi-omics data. BMC Bioinform 23(1)
11. Qazi S, Khanam A, Raza K (2023) Potential role of the nature-inspired algorithms for classification of high-dimensional and complex gene expression data. In: Raza K (ed) Nature-inspired intelligent computing techniques in bioinformatics. Springer Nature Singapore, Singapore, pp 89–102
12. Qazi W et al (2023) The scope and applications of nature-inspired computing in bioinformatics. In: Raza K (ed) Nature-inspired intelligent computing techniques in bioinformatics. Springer Nature Singapore, Singapore, pp 3–18
13. Dimitrakopoulos C et al (2018) Network-based integration of multi-omics data for prioritizing cancer genes. Bioinformatics 34(14):2441–2448
14. Yousef M, Kumar A, Bakir-Gungor B (2020) Application of biological domain knowledge based feature selection on gene expression data. Entropy 23(1):2–2
15. Coleto-Alcudia V, Vega-Rodríguez MA (2020) Artificial Bee Colony algorithm based on dominance (ABCD) for a hybrid gene selection method. Knowl-Based Syst 205:106323–106323
16. Krassowski M et al (2020) State of the field in multi-omics research: from computational needs to data mining and sharing. Front Genet 11
17. Subramanian I et al (2020) Multi-omics data integration, interpretation, and its application. Bioinform Biol Insights 14
18. Athieniti E, Spyrou GM (2023) A guide to multi-omics data collection and integration for translational medicine. Comput Struct Biotechnol J 21:134–149
19. Hasin Y, Seldin M, Lusis A (2017) Multi-omics approaches to disease. Genome Biol 18(1)
20. Wang B et al (2014) Similarity network fusion for aggregating data types on a genomic scale. Nat Methods 11(3):333–337
21. Lock EF et al (2013) Joint and individual variation explained (JIVE) for integrated analysis of multiple data types. Annals Appl Stat **7**(1)
22. Sathyanarayanan A et al (2019) A comparative study of multi-omics integration tools for cancer driver gene identification and tumour subtyping. Brief Bioinform 21(6):1920–1936
23. Vaske CJ et al (2010) Inference of patient-specific pathway activities from multi-dimensional cancer genomics data using PARADIGM. Bioinformatics (Oxford, England) 26(12):i237-245
24. Zhang S et al (2012) Discovery of multi-dimensional modules by integrative analysis of cancer genomic data. Nucleic Acids Res 40(19):9379–9391
25. Yang Z, Michailidis G (2015) A non-negative matrix factorization method for detecting modules in heterogeneous omics multi-modal data. Bioinformatics: btv544

26. Mo Q et al (2013) Pattern discovery and cancer gene identification in integrated cancer genomic data. Proc Natl Acad Sci 110(11):4245–4250
27. Mo Q et al (2017) A fully Bayesian latent variable model for integrative clustering analysis of multi-type omics data. Biostatistics 19(1):71–86
28. Meng C et al (2015) MoCluster: identifying joint patterns across multiple omics data sets. J Proteome Res 15(3):755–765
29. Argelaguet R et al (2018) Multi-omics factor analysis—a framework for unsupervised integration of multi-omics data sets. Mol Syst Biol 14(6)
30. Lock EF, Dunson DB (2013) Bayesian consensus clustering. Bioinformatics 29(20):2610–2616
31. Kirk P et al (2012) Bayesian correlated clustering to integrate multiple datasets. Bioinformatics 28(24):3290–3297
32. Wang W et al (2012) IBAG: integrative Bayesian analysis of high-dimensional multiplatform genomics data. Bioinformatics 29(2):149–159
33. Koh HWL et al (2019) IOmicsPASS: network-based integration of multiomics data for predictive subnetwork discovery. NPJ Syst Biol Appl 5:22–22
34. Rappoport N, Shamir R (2018) Multi-omic and multi-view clustering algorithms: review and cancer benchmark. Nucleic Acids Res 46(20):10546–10562
35. Shi Q et al (2017) Pattern fusion analysis by adaptive alignment of multiple heterogeneous omics data 33(17):2706–2714
36. El-Manzalawy Y (2018) CCA-based multi-view feature selection for multi-omics data integration
37. Wu D et al (2015) Fast dimension reduction and integrative clustering of multi-omics data using low-rank approximation: application to cancer molecular classification. BMC Genomics 16(1)
38. Hoadley KA et al (2014) Multiplatform analysis of 12 cancer types reveals molecular classification within and across tissues of origin. Cell 158(4):929–944
39. Bonnet E, Calzone L, Michoel T (2015) Integrative multi-omics module network inference with lemon-tree. PLoS Comput Biol 11(2):e1003983–e1003983
40. Yang B et al (2018) Deep subspace similarity fusion for the prediction of cancer subtypes
41. Ma T, Zhang A (2018) Affinity network fusion and semi-supervised learning for cancer patient clustering. Methods 145:16–24
42. Lim SY et al (2022) Multi-omics investigation into acute myocardial infarction: an integrative method revealing interconnections amongst the metabolome, lipidome, glycome, and metallome. Metabolites 12(11)
43. Zhan C et al (2023) From multi-omics approaches to personalized medicine in myocardial infarction. Front Cardiovasc Med 10:1250340
44. Karaboga D, Basturk B (2005) Artificial Bee Colony (ABC) optimization algorithm for solving constrained optimization problems. Lecture Notes in Computer Science, pp 789–798
45. Bonabeau E, Dorigo M, Theraulaz G (1999) Swarm intelligence. Oxford University Press
46. Karaboga D, Akay B (2009) A comparative study of Artificial Bee Colony algorithm. Appl Math Comput 214(1):108–132
47. Aziz R, Verma CK, Srivastava N (2017) A novel approach for dimension reduction of microarray. Comput Biol Chem 71:161–169
48. Alshamlan H, Badr G, Alohali Y (2015) MRMR-ABC: a hybrid gene selection algorithm for cancer classification using microarray gene expression profiling. Biomed Res Int 2015:1–15
49. Kaur P, Singh A, Chana I (2022) BSense: a parallel Bayesian hyperparameter optimized stacked ensemble model for breast cancer survival prediction. J Comput Sci 60:101570–101570
50. Wu L et al (2022) Integrated multi-omics for novel aging biomarkers and antiaging targets. Biomolecules 12(1):39–39
51. Pan Y, Lei X, Zhang Y-C (2021) Association predictions of genomics, proteinomics, transcriptomics, microbiome, metabolomics, pathomics, radiomics, drug, symptoms, environment factor, and disease networks: a comprehensive approach 42(1):441–461
52. Banerjee S, Prabhu Basrur N, Rai PS (2021) Omics technologies in personalized combination therapy for cardiovascular diseases: challenges and opportunities. Pers Med

53. Hussain SF, Pervez A, Hussain M (2020) Co-clustering optimization using Artificial Bee Colony (ABC) algorithm. Appl Soft Comput 97:106725–106725
54. Akbari R et al (2012) A multi-objective artificial bee colony algorithm. Swarm Evol Comput 2:39–52
55. Bai L et al (2022) Advances and applications of machine learning and intelligent optimization algorithms in genome-scale metabolic network models. Syst Microbiol Biomanuf
56. Mostafa RR et al (2024) An adaptive hybrid mutated differential evolution feature selection method for low and high-dimensional medical datasets. Knowl-Based Syst 283:111218–111218
57. Adossa N et al (2021) Computational strategies for single-cell multi-omics integration. Comput Struct Biotechnol J 19:2588–2596
58. Forsati R et al (2012) A novel approach for feature selection based on the Bee Colony optimization. Int J Comput Appl 43:13–16
59. Forghany Z, Forghany MD (2012) Gene regulatory network model identification using artificial bee colony and swarm intelligence. In: IEEE congress on evolutionary computation (CEC), pp 1–6
60. Aziz RM et al (2022) Cuckoo search-based optimization for cancer classification: a new hybrid approach. J Comput Biol 29(6):565–584

The manufacturer's authorised representative in the EU is Springer Nature Customer Service Centre GmbH, Europaplatz 3, 69115 Heidelberg, Germany. If you have any concerns regarding our products, please contact ProductSafety@springernature.com

Printed and bound by CPI Group (UK) Ltd, Croydon, CR0 4YY
01/10/2026
02225969-0001